Annals of Mathematics Studies

Number 67

PROFINITE GROUPS, ARITHMETIC, AND GEOMETRY

BY

STEPHEN S. SHATZ

PRINCETON UNIVERSITY PRESS

AND

UNIVERSITY OF TOKYO PRESS

1972

LC Card: 77-126832
ISBN: 0-691-08017-8
AMS 1968: Primary—2050, 2051, 1067, 1068
Secondary—1015, 1077, 1078,
1435, 1440, 1810, 1815, 1820, 5532

Published in Japan exclusively by
University of Tokyo Press;
in other parts of the world by
Princeton University Press

Printed in the United States of America

For Marilyn, Geoffrey, and Adria

PREFACE

This short volume reproduces in its first five chapters the essential content of a one semester course I gave at the University of Pennsylvania in the Spring of 1968. The last chapter covers material I had hoped to include, but could not for lack of time. My aim in giving such a course was to give students a body of material upon which some of the modern research in Diophantine geometry and higher arithmetic is based, and to do this in a way which emphasized the many interesting roads out of these elementary foundations. I wanted to enable them to start reading the literature in those areas of mathematics I personally find very interesting. The enthusiastic reception of my able class convinced me that this material was deserving of a larger audience, and I resolved to write up the notes for the course.

Anyone conversant with this subject will note how indebted I am to many authors. Foremost among them is J. P. Serre from whose works I have taken a great deal,* and there is John Tate whose unpublished work and unmistakeable stamp appear here in many places. Responsibility for any errors is, of course, my own.

The highly personal nature of these notes has not blinded me to their many faults, notably the lack of good examples and exercises, and the uneven tempo of exposition. In attempting to retain the flavor of the original lectures, I have allowed these and other faults to stand; I hope they do not vitiate the aim of the project. I have also included a large but not exhaustive bibliography; it covers the main references, and the bibliographies in these works and will serve as a good guide to the literature.

* Especially from [SCL, SG] where the smooth treatment is impossible to improve upon.

CONTENTS

PROFINITE GROUPS,

ARITHMETIC, AND GEOMETRY

CHAPTER I

PROFINITE GROUPS

§0. Preliminaries on projective and inductive limits*

Let Λ be a set (the index set) and suppose S_α, for $\alpha \in \Lambda$ is a family of sets (groups, rings, etc.) indexed by Λ. We assume Λ is partially ordered (by $\alpha \leqq \beta$) and satisfies the

$$\text{Moore-Smith Property: } (\forall \alpha, \beta)(\exists \gamma)(\alpha \leqq \gamma \text{ and } \beta \leqq \gamma) \ .$$

Now we assume given the following data:

(1) $$(\forall \alpha, \beta)(\alpha \leqq \beta \Longrightarrow \exists \phi_\alpha^\beta : S_\alpha \to S_\beta) \ ,$$

and the maps (group-homomorphisms, ring homomorphisms, etc.) are *consistent* in the sense that: given α, β, γ with $\alpha \leqq \beta \leqq \gamma$, then the induced diagram

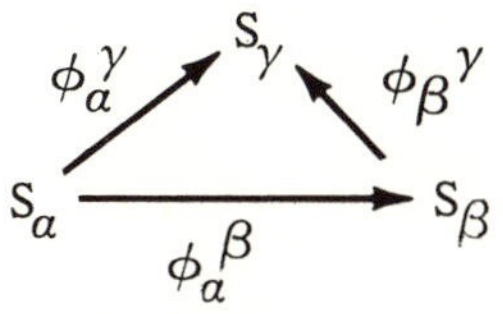

commutes. Furthermore, we insist that $\phi_\alpha^\alpha = \text{id}$, all α.

Any system, as above, will be called a *direct mapping family* of sets (groups, rings), and denoted $\{S_\alpha, \phi_\alpha^\beta\}$. From now on, we will concentrate on the set-theoretic formulation, leaving to the reader the task of formulating all definitions and results for groups, rings, or objects of other categories.

* Only the concept of lim based on a directed index set, not on the more general notion of "index category," is introduced here.

If instead of the equation (1), we write

$$(\forall \alpha, \beta)(\alpha \leqq \beta \Longrightarrow \exists \phi_\beta^\alpha : S_\beta \to S_\alpha) \tag{2}$$

and demand the obvious consistency, we get a *projective (or inverse) mapping family* of sets (etc.) —this time with the notation

$$\{S_\alpha, \phi_\beta^\alpha\} .$$

Examples. (1D) $S_\alpha = Z$, $\Lambda =$ natural numbers: $1, 2, \ldots$, and the ordering on Λ is by cardinality (i.e., the usual ordering). $\phi_n^{n+1}: Z \to Z$ is "multiplication by p," where p is a fixed prime number.

(1P) Same as (1D), only in this case ϕ_{n+1}^n is multiplication by p.

(2D) $\Lambda =$ natural numbers, partially ordered by division. $S_n = Z/nZ$, and if $n \leqq m$ (i.e., $n|m$), then ϕ_n^m is multiplication by m/n.

(2P) Same S_n and Λ as (2D). This time ϕ_m^n is the natural projection $Z/mZ \to Z/nZ$.

(3D) G is a group; Λ is the family of all finite subgroups of G, partially ordered by inclusion. S_α is α considered as subgroup of G, not as element of Λ. The map ϕ_α^β is the natural inclusion $S_\alpha \to S_\beta$.*

(3P) G is a group, Λ is the family of all normal subgroups of G of finite index. If α is in Λ, S_α is G/α, a finite group. Partially order Λ *via*: $\alpha \leqq \beta \Longleftrightarrow \beta \subseteq \alpha$ as subgroups; let ϕ_β^α be the natural projection $S_\beta \to S_\alpha$. (Observe (3P) generalizes (2P).)

If T is a set, and if $\psi_\alpha: S_\alpha \to T$ are given maps, one for each $\alpha \in \Lambda$, which are consistent with the ϕ_α^β in the sense that the diagram

(3)

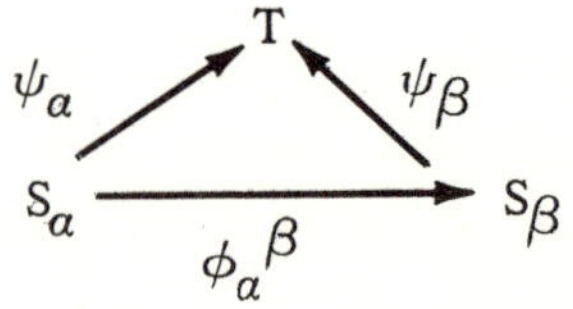

commutes for all $\alpha \leqq \beta$; then we may inquire as to how "faithfully $\{T, \psi_\alpha\}$ possesses all the information contained in $\{S_\alpha, \phi_\alpha^\beta\}$"? More exactly,

* We assume G is abelian.

we may ask: Does there exist a "universal" object $\{T, \psi_\alpha\}_{\alpha \in \Lambda}$ from which the others may be recaptured? (For projective systems the ψ_α take T to S_α for all α, and the diagram is changed *mutatis mutandis*.) As is usual in such circumstances, what we are asking for is the *representability of a certain functor* —or, in older terminology, the *solution of a universal mapping problem.* The functor we wish to represent is F where

$$F(T) = \{\text{families } \{\psi_\alpha\}_{\alpha \in \Lambda} \mid \psi_\alpha : S_\alpha \to T \text{ and}$$
$$\text{diagram (3) commutes for all } \alpha \leqq \beta\} .$$

The representing object will then be a set S and a "universal" family $\{\phi_\alpha\}_{\alpha \in \Lambda}$, where $\phi_\alpha : S_\alpha \to S$ and the ϕ_α are consistent. In terms of universal mapping properties, the object we seek will have the

Direct universal mapping property (DUMP):

(a) $\{S, \phi_\alpha\}_{\alpha \in \Lambda}$ is a set and a consistent family indexed by Λ; $\phi_\alpha : S_\alpha \to S$ and

(b) Given any other pair $\{T, \psi_\alpha\}_{\alpha \in \Lambda}$ having (a), *there exists a unique* map

$$\phi : S \to T$$

so that the diagram

(4)

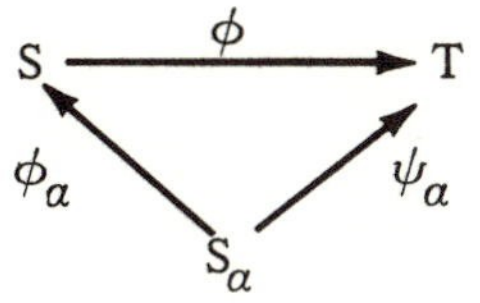

commutes for every $\alpha \in \Lambda$.

(The reader may formulate the analogous universal mapping problem for inverse families by turning arrows around and so arrive at a formulation of the *Projective universal mapping property* (PUMP).)

Again, as is usual, if the objects we seek exist then they will be unique up to unique isomorphism. Any object satisfying the DUMP is called the *direct limit* of the family $\{S_\alpha, \phi_\alpha^\beta\}$ —by abuse of notation it is written

$$S = \operatorname*{dir\,lim}_\alpha S_\alpha .$$

Similarly, any object satisfying the PUMP is the *projective limit* of the family $\{S_\alpha, \phi_\beta^\alpha\}$ —written

$$S = \operatorname*{proj\,lim}_\alpha S_\alpha .$$

For our purposes, the following theorem is sufficient.

THEOREM 1. *In the categories of sets, groups, rings, and topological groups, direct and inverse limits always exist.*

Proof: We content ourselves with a sketch—and that only for the case of sets. The other categories are entirely similar and obvious modifications of our proof yield the desired results.

Let $\{S_\alpha, \phi_\alpha^\beta\}$ be a direct mapping family of sets. We may and do assume that the S_α are disjoint, and we form their union $\mathcal{S} = \bigcup_\alpha S_\alpha$. In $\mathcal{S}$ introduce the relation, $\sim$, by:

If $x \in \mathcal{S}$, $y \in \mathcal{S}$, (say $x \in S_\alpha$, $y \in S_\beta$) then $x \sim y$ if and only if $(\exists\gamma)(\alpha \leqq \gamma, \beta \leqq \gamma)$ such that $\phi_a^\gamma(x) = \phi_\beta^\gamma(y)$.

One sees trivially that $\sim$ is an equivalence relation, and we may form $S = \mathcal{S}/\sim$. Let $\phi_\alpha \colon S_\alpha \to S$ be the composition: $S_\alpha \to \mathcal{S} \to \mathcal{S}/\sim$. It is easily checked that $\{S, \phi_\alpha\}_{\alpha\in\Lambda}$ has the DUMP.

Now let $\{S_\alpha, \phi_\beta^\alpha\}$ be an inverse mapping family of sets. Form the cartesian product $X = \prod_\alpha S_\alpha$. Let S be the subset of X consisting of all *consistent* tuples (x_α) i.e.,

$$(x_\alpha) \in S \iff (\forall\beta,\gamma)(\beta \leqq \gamma \implies \phi_\gamma^\beta(x_\gamma) = x_\beta) .$$

Let ϕ_α be the αth projection restricted to S. One checks that $\{S, \phi_\alpha\}_{\alpha\in\Lambda}$ has the PUMP. (Note: It may happen that $S = \emptyset$.) Q.E.D.

Examples. These examples refer to the earlier ones on page 4.

(1D) $\operatorname{dir}\lim_n \{S_n, \phi_n^{n+1}\} = \{\frac{r}{p^s} \mid r, s \in \mathbf{Z}\}$

(1P) $\operatorname{proj}\lim \{S_n, \phi_{n+1}^{n}\} = (0)$

(2D) $\operatorname{dir}\lim \{S_n, \phi_n^{m}\} = \mathbf{Q}/\mathbf{Z}$

(2P) $\operatorname{proj}\lim \{S_n, \phi_m^{n}\} = \hat{\mathbf{Z}} =$ "completion of $\mathbf{Z}$ in the ideal topology"

$\cong \Pi_p \mathbf{Z}_p$, where $\mathbf{Z}_p$ is the ring of p-adic integers.

(3 D) If G is abelian, $\operatorname{dir}\lim_\alpha S_\alpha$ is the torsion subgroup of G.

(3 P) $\operatorname{proj}\lim_\alpha S_\alpha$ is called the "*strong*" *completion* of G.

§1. Profinite Groups

DEFINITION 1. A *profinite group* is a projective limit of finite groups.

Observe that examples (2P) and (3P) are profinite groups. From the proof of Theorem 1, since every finite group is a compact, Hausdorff group, one sees that *the projective limit of finite groups is, in a natural way, a compact Hausdorff group.* The following theorem specifies how we may isolate the profinite groups from among all compact groups.

THEOREM 2. *The following conditions are equivalent.*

(1) G *is a profinite group.*

(2) G *is a compact, Hausdorff group in which the family of open normal subgroups forms a fundamental system of neighborhoods at* 1.

(3) G *is a compact, totally disconnected, Hausdorff group.*

Proof: (1) $\Longrightarrow$ (2). Since G is profinite, there are finite groups G_α and maps $\phi_\beta^\alpha : G_\beta \to G_\alpha$ such that $G = \operatorname{proj}\lim_\alpha G_\alpha$. Hence, there are maps $\phi_\alpha : G \to G_\alpha$, for all α. Let $U_\alpha = \ker \phi_\alpha$, then the U_α are open, normal subgroups of G. Clearly, (2) will hold if we succeed in proving that $\bigcap_\alpha U_\alpha = \{1\}$. But, if $\xi \in U_\alpha$ for every α, then using the explicit form of G gotten from the proof of Theorem 1, one sees that ξ as a tuple has α^{th} projection 1 everywhere. It follows immediately that $\xi = 1$, as required.

(2) $\Longrightarrow$ (3). Let N be the connected component of identity in G. If $\{U_\alpha\}$ denotes the family of all open, normal subgroups of G, let $N_\alpha = U_\alpha \cap N$ for all α. The groups N_α are open, normal subgroups of N; as N is connected, we must have $N_\alpha = N$ for all α. But then

$$N = \bigcap_\alpha N_\alpha = \bigcap_\alpha N \cap U_\alpha = N \cap \bigcap_\alpha U_\alpha = N \cap \{1\} .$$

(The last equality by hypothesis (2).) Thus $N = \{1\}$, and (3) is proved.

(3) $\Longrightarrow$ (1). It follows from our hypothesis, and fromTheorem 17, Chapter III of [P], that given any open neighborhood U of 1 in G, there exists an open subgroup H of G with $H \subseteq U$. Since G is compact and H is open, the index $(G:H)$ is finite, and therefore $(G: N_G(H)) < \infty$. From this, it follows that the distinct conjugates xHx^{-1} of H in G are finite in number, so their intersection K forms an open, normal subgroup of G contained in H; hence, in U. As U was an arbitrary open neighborhood of 1, we see that the intersection of all open, normal subgroups of G reduces to the identity. Let $\{U_\alpha\}$ denote the family of all open, normal subgroups of G (so that we have shown $\bigcap_\alpha U_\alpha = \{1\}$). The groups $G/U_\alpha = G_\alpha$ form in a natural way a projective mapping family of *finite* groups. Our theorem now follows from the more general

LEMMA 1. *Let* G *be a compact, Hausdorff group; let* G_α, S_α *be two families of closed subgroups of* G, *indexed by the directed set* Λ. *Assume that* $S_\alpha \lhd G_\alpha$ *for all* α; *and that for* $\beta \geqq \alpha$, $G_\alpha \supseteq G_\beta$ *and* $S_\alpha \supseteq S_\beta$. *Then the groups* $H_\alpha = G_\alpha/S_\alpha$ *form a projective mapping family in a natural way, and*

$$\operatorname*{proj\,lim}_\alpha H_\alpha = \operatorname*{proj\,lim}_\alpha G_\alpha/S_\alpha = \bigcap_\alpha G_\alpha/\bigcap_\alpha S_\alpha .$$

Proof: If $\beta \geqq \alpha$, then $S_\beta \subseteq S_\alpha$; so one has the maps

$$\phi_\beta{}^\alpha : H_\beta = G_\beta / S_\beta \to G_\alpha / S_\alpha = H_\alpha$$

given by $\phi_\beta{}^\alpha(xS_\beta) = xS_\alpha$. The consistency of these maps is trivial. Let $\Gamma = \operatorname{proj\,lim}_\alpha H_\alpha$, and let $\phi_\alpha : \Gamma \to H_\alpha$ be the canonical homomorphisms. There are obvious maps $\psi_\alpha : \bigcap G_\beta / \bigcap S_\beta = H \to H_\alpha$, for each α and they are consistent with the $\phi_\beta{}^\alpha$. Hence, by the PUMP, we obtain a map

$$\phi : H = \bigcap G_\alpha / \bigcap S_\alpha \to \Gamma \ .$$

Here, map means continuous group homomorphism, so when ϕ is proven bijective the compact Hausdorff property of the groups in question will show that ϕ is automatically a homeomorphism.

Let $\xi \in H$, and assume $\phi(\xi) = 1 \in \Gamma$. The diagram

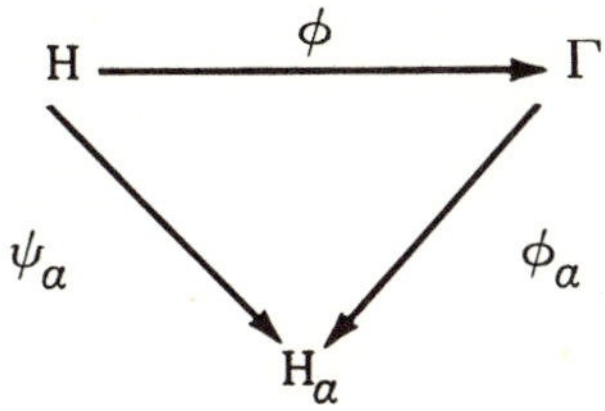

shows that $\psi_\alpha(\xi) = 1$ for all α. Thus $\xi \in S_\alpha$ for all α; hence, ξ represents 1 in H and ϕ is injective. Given $x \in \Gamma$, let $x_\alpha = \phi_\alpha(x) \in H_\alpha$ for each α. As H_α is G_α / S_α, there exists a $\xi_\alpha \in G_\alpha$ such that ξ_α mod S_α is x_α. The net $\{\xi_\alpha\}_\alpha$ possesses a converging subnet $\{\xi_{\alpha_\nu}\}_\nu$ because all the ξ_α lie in G and G is compact. Let $\tilde{\xi}$ be the limit of this subnet. Now given any α, the set of all α_ν with $\alpha_\nu \geqq \alpha$ is final in the α_ν; hence, $\lim_{\alpha_\nu \geqq \alpha} \xi_{\alpha_\nu} = \tilde{\xi}$. But $\xi_{\alpha_\nu} \in G_{\alpha_\nu} \subseteq G_\alpha$ if $\alpha_\nu \geqq \alpha$, and as G_α is closed in G, we deduce that $\tilde{\xi} \in G_\alpha$. Thus $\tilde{\xi} \in \bigcap_\alpha G_\alpha$. Let ξ be the coset of $\tilde{\xi}$ mod $\bigcap_\alpha S_\alpha$, i.e., $\xi \in H$. All we need now show is $\psi_\alpha(\tilde{\xi}) = x_\alpha$ for every α. This is very easy as follows: The x_α are consistent; hence, for all $\beta \geqq \alpha$, ξ_β mod $S_\alpha = \xi_\alpha$ mod S_α. Given any α, for any $\beta \geqq \alpha$, we can find

$s_{\alpha\beta} \in S_\alpha$ with $\xi_\beta = \xi_\alpha s_{\alpha\beta}$. The $\{s_{\alpha\beta}\}_\beta$ form a net in S_α, so there is a converging subnet. Look at the subnet of this newly chosen subnet formed by those β beyond some number a_ν of our previous indexing set. We have

$$\lim_{\substack{\beta \text{ in new} \\ \text{family}}} s_{\alpha\beta} = s_\alpha \in S_\alpha,$$

and hence,

$$\tilde{\xi} = \lim_\beta \xi_\beta = \lim_\beta \xi_\alpha s_{\alpha\beta} = \xi_\alpha s_\alpha .$$

The last equation when unscrambled in terms of $\tilde{\xi}$ and ψ_α says precisely $\psi_\alpha(\tilde{\xi}) = x_\alpha$. Q.E.D.

Here are some facts concerning profinite groups whose proofs are easy exercises.

(Fact 1) Every closed subgroup and every quotient group of a profinite group is profinite.

(Fact 2) The product or projective limit of a family of profinite groups is profinite.

(Fact 3) If S is any subgroup of a profinite group G, then its closure $\overline{S}$ is given by

$$\overline{S} = \bigcap \{U \mid U \text{ is an open normal subgroup of } G \text{ and } S \subseteq U\}.$$

(Fact 4) If G is an abelian group, then G is profinite if and only if its Pontrjagin dual is a torsion group. In this way, the category of profinite abelian groups is the dual of the category of torsion abelian groups.

An important theorem for the cohomology theory of profinite groups is:

THEOREM 3. (Cross-section Theorem) *Let* G *be a profinite group and let* N *be a closed normal subgroup. Then there exists a continuous cross-section for the map* $\pi : G \to G/N$. *That is, there is a continuous map* $\phi : G/N \to G$ *(not a homomorphism in general) such that* $\phi(x) \in x$ *or, equivalently,* $\pi(\phi(x)) = x$.

Proof: Let $\mathcal{S}$ be the set of all *closed* normal subgroups S of G such that $S \subseteq N$. Consider pairs $<S, \lambda>$ where $\lambda: G/N \to G/S$ is a cross-section (continuous of course) of the projection $G/S \to G/N$. Call the set of such pairs $\mathfrak{P}$ and partially order $\mathfrak{P}$ in the usual fashion, viz:

$$<S, \lambda> \leqq <T, \mu> \iff (1)\ T \subseteq S \text{ and (2) the diagram}$$

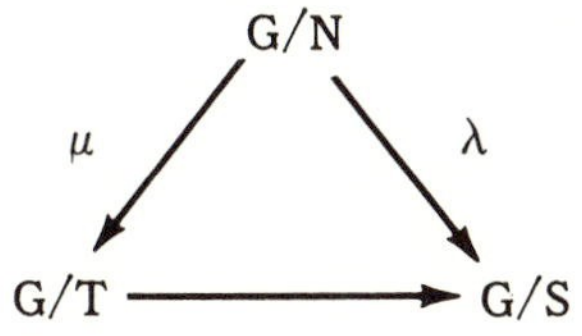

commutes (i.e., $\lambda(x) = S\mu(x)$).

Now $\mathfrak{P} \neq \emptyset$ as $<N, id> \in \mathfrak{P}$. If $<S_\alpha, \lambda_\alpha>$ is a totally ordered subfamily of $\mathfrak{P}$, let $S = \bigcap_\alpha S_\alpha$. By Lemma 1, $G/S = \operatorname{proj}\lim_\alpha G/S_\alpha$; so, as $\lambda_\beta(x) \to \lambda_\alpha(x)$ for $\beta \geqq \alpha$ under the map $G/S_\beta \to G/S_\alpha$, we see that for each $x \in G/N$, the family $\{\lambda_\alpha(x)\}$ represents a point $\lambda(x) \in G/S$. Clearly, $<S, \lambda>$ belongs to $\mathfrak{P}$, and we have proved $\mathfrak{P}$ is inductive.

By Zorn's lemma, there exists a maximal pair $<S, \lambda>$ in $\mathfrak{P}$. It is only necessary to prove that $S = \{1\}$. Were $S \neq \{1\}$, there would exist an open normal subgroup U of G with $S \not\subseteq U$. Therefore, $S \cap U < S$. We shall construct the section $G/S \to G/S \cap U$, and this will yield the desired contradiction.

Let $\underline{G} = G/S \cap U$, $\underline{S} = S/S \cap U$, then G/S is $\underline{G}/\underline{S}$ and our problem is to construct the section $\underline{G}/\underline{S} \to \underline{G}$ for the profinite group $\underline{G}$. But $\underline{S}$ is isomorphic to SU/U which is *finite* (being contained in G/U). Hence, our problem is reduced to the case of a finite normal subgroup. As $\underline{S}$ is finite, there is an open normal subgroup $\underline{V}$ of $\underline{G}$ such that $\underline{S} \cap \underline{V} = \{1\}$. The projection $\underline{G} \to \underline{G}/\underline{S}$ when restricted to $\underline{V}$ is a monomorphism. Since all groups are compact Hausdorff, $\underline{V}$ is topologically isomorphic to its image in $\underline{G}/\underline{S}$. The inverse map of this isomorphism is a continuous map from an *open* subgroup of $\underline{G}/\underline{S}$ onto an open subgroup of $\underline{G}$. Now a translation of this mapping to the finitely many cosets in $\underline{G}/\underline{S}$ yields our section $\underline{G}/\underline{S} \to \underline{G}$. Q.E.D.

COROLLARY. *If* M, N *are closed, normal subgroups of the profinite group* G, *with* $N \supseteq M$, *then there exists a section* $G/N \to G/M$.

Remark. One of the features of the theory of profinite groups which makes it important for various problems in number theory and geometry is that, in large measure, many mapping properties and number-theoretic properties of finite groups are (when attention is paid to their proper formulation) in reality valid for the larger class of profinite groups. An example is the cross-section theorem above; others follow in the next section.

§2. Supernatural numbers and the Sylow Theory.

DEFINITION 2. *A supernatural number* is a formal product

$$\prod_p p^{n_p}$$

where the product is taken over *all* prime numbers and n_p is an integer $0 \leqq n_p \leqq \infty$ (observe: ∞ is allowed!).

The theory of such numbers is entirely multiplicative, and it is easy to define the product, l.c.m. and g.c.d. of supernatural numbers or s-numbers as we shall call them. The connection with profinite groups is essentially given in the following definition.

DEFINITION 3. Let G be a profinite group, H a closed subgroup. By the *index of* H *in* G, we mean the s-number given by

$$(G:H) = \text{l.c.m.}(G/U : H/H \cap U) \text{ as } U \text{ ranges over all open normal subgroups of } G.$$

The *order of* G is defined by the equation

$$|G| = (G:1) = \underset{U}{\text{l.c.m.}} |G/U| .$$

Exercise. $(G:H) = \text{l.c.m.}(G:U)$ over those open, normal U with $H \subseteq U$.

If s is an s-number, then s is a p-*power* (*prime—to—*p) $\Longleftrightarrow$ $n_q = 0$ for all $q \neq p$ ($n_p = 0$). G is a p-*group* $\Longleftrightarrow$ $|G|$ is a p-power.

The first indication that things are all right is

PROPOSITION 1. (Lagrange's Theorem). *Let* $S \supseteq K$ *be closed subgroups of a profinite group* G. *Then*

$$(G:K) = (G:S)(S:K)$$

Proof: $(G:K) = \underset{\alpha}{\text{l.c.m.}}\,(G/U_\alpha : K/K \cap U_\alpha)$, where the U_α range over all open, normal subgroups of G. Now $K/K \cap U_\alpha = KU_\alpha / U_\alpha$ and $SU_\alpha \supseteq KU_\alpha$. Hence,

$$\begin{aligned}(G/U_\alpha : K/K \cap U_\alpha) &= (G/U_\alpha : SU_\alpha/U_\alpha)(SU_\alpha/U_\alpha : KU_\alpha/U_\alpha)\\ &= (G:SU_\alpha)(S/S \cap U_\alpha : K/K \cap U_\alpha)\ .\end{aligned}$$

However, $S \cap U_\alpha = S_\alpha$ is open, normal in S, and the family of such is final in the set of all open, normal subgroups of S. Therefore,

$$(G/U_\alpha : K/K \cap U_\alpha) = (G:SU_\alpha)(S/S_\alpha : K/K \cap S_\alpha)\ .$$

It is easy to see (but requires a small separate argument which is left to the reader) that

$$(G:K) = \underset{\alpha}{\text{l.c.m.}}(G:SU_\alpha)\ \underset{\alpha}{\text{l.c.m.}}\,(S/S_\alpha : K/K \cap S_\alpha)\ ;$$

hence, $(G:K) = (G:S)(S:K)$. Q.E.D.

A p-*Sylow group* S of a profinite group G is a closed sub-p-group such that $(G:S)$ is prime—to—p.

THEOREM 4. (Sylow Theorem for Profinite Groups). *Let* G *be a profinite group, and let* p *be a prime number. Then*

(1) G *possesses* p-*Sylow subgroups;*

(2) *If* T *is any* p-*subgroup of* G *then* T *is contained in some* p-*Sylow subgroup of* G.

(3) *Any two* p*-Sylow subgroups of* G *are conjugate in* G.

Proof: We shall give two proofs of part (1) of this theorem. The first is due to John Tate, the second to J.-P. Serre. For the first, we let $\mathcal{S}$ be the set of all subgroups $H \subseteq G$ such that $(G:H)$ is prime-to-p. Since $G \in \mathcal{S}$, $\mathcal{S}$ is non-empty and we partially order it by inclusion. If $\{H_\alpha\}$ is a chain in $\mathcal{S}$, let $H = \bigcap_\alpha H_\alpha$. Let U be open, normal in G, then $HU \supseteq H$ and is open in G. Because the H_α are compact, there exists an index α for which $H_\alpha \subseteq HU$. Thus $(G:HU)$ divides $(G:H_\alpha)$; hence, $(G:HU)$ is prime-to-p. It follows that $H \in \mathcal{S}$; whence, $\mathcal{S}$ is inductive. Choose a minimal element, P, of $\mathcal{S}$. I claim $|P|$ is a p-power. If not, there would exist an open, normal subgroup U of G with $(P:P \cap U)$ not a p-power. As $P/P \cap U$ is finite, it possesses a p-Sylow subgroup $Q/P \cap U$, with $P \neq Q$. But then, $(P:Q)$ is prime-to-p, and Q is smaller than P in $\mathcal{S}$, a contradiction.

The second (Serre's) proof of (1) uses the following well-known fact: *The projective limit of a non-empty family of finite sets is non-empty.* This being said, let $\{U_\alpha\}$ be the family of all open, normal subgroups of G, and let $P(U_\alpha)$ be the set of all p-Sylow subgroups of G/U_α. By the ordinary Sylow Theorem, the sets $P(U_\alpha)$ are non-empty and finite. Moreover, if $\beta \geqq \alpha$, then $U_\beta \subseteq U_\alpha$, so the surjection $G/U_\beta \to G/U_\alpha$ yields a map $P(U_\beta) \to P(U_\alpha)$. By our remark, there exists an element $\mathfrak{P}$ in $\operatorname{proj\,lim}_\alpha P(U_\alpha)$. What is such a $\mathfrak{P}$? It is a family P_α of subgroups of G/U_α such that each P_α is a p-Sylow subgroup of G/U_α and such that for $\beta \geqq \alpha$, $P_\beta \to P_\alpha$ exists and is consistent for $\gamma \geqq \beta \geqq \alpha$, In other words, $\mathfrak{P}$ is a projective mapping family of p-Sylow sugroups from G/U_α, for each α. Let $P = \operatorname{proj\,lim}_\alpha P_\alpha$, then P is a p-group. I claim $(G:P)$ is prime-to-p. But,

$$(G:P) = \text{l.c.m.}(G/U_\alpha : P/P \cap U_\alpha) = \text{l.c.m.}(G/U_\alpha : P_\alpha)$$

$$= \text{prime-to-p} .$$

Finally, to prove (2) and (3) it suffices to prove: *If* T *is a* p*-subgroup* *of* G, *there is some* $\sigma \in G$ *with* $T^\sigma \subseteq P$, *where* P *is a given* p-*Sylow subgroup of* G. Here, the notation T^σ means $\sigma^{-1}T\sigma$. Choose any open normal subgroup U of G, and consider TU and PU. We know that PU/U is a p-Sylow subgroup of G/U, so by finite Sylow Theory, there exists a $\bar{\sigma}_U$ in G/U with

$$(*) \qquad (TU/U)^{\bar{\sigma}_U} \subseteq (PU/U) \ .$$

Let R_U be the set of all σ_U in G such that $\bar{\sigma}_U$ satisfies (*) above. Then (1) $R_U \neq \emptyset$ for any U, (2) R_U is a union of cosets of U in G so is closed in G, and (3) the sets R_U have the finite intersection property. (The last assertion follows from $R_V \subseteq R_U$ if $V \subseteq U$.) By the compactness of G, the intersection $\bigcap_U R_U$ is non-empty; and it is clear that $\sigma \in \bigcap_U R_U$ is the required element of G. Q.E.D.

Examples. (4) If G is a group generated by a set $\{x_\alpha\}$ of elements, let $\mathcal{S}$ be the set of all normal subgroups U of G such that (a) G/U is finite, (b) almost all x_α lie in U. Let $\hat{G} = \operatorname{proj\,lim}_U G/U$, then $\hat{G}$ is called the *profinite completion of* G. For example if G is Z we get $\hat{Z}$ as in example (2P).

(5) Let G = F(X) be the free-group on X. Its profinite completion is called the *free-profinite group on* X. If we restrict attention to those subgroups U for which in addition (c) (G: U) is a p-power, then we get $F_p(X)$ (or in example (4) $\hat{G}_p$) which is called the *free-profinite* p-*group on* X (or the p-*profinite completion* of G). If X is the set with one element, we have F(X) = Z, $F_p(X) = Z_p$ = p-adic integers (see example (2P)) = p-Sylow subgroup of $\hat{Z}$.

CHAPTER II

COHOMOLOGY OF PROFINITE GROUPS

§1. δ-functors and the definition of the cohomology of profinite groups.

Let $\mathcal{A}$ be an abelian category (think of modules over a ring, or abelian groups on which some group acts *via* automorphisms), then we have the notion of an exact sequence of objects of $\mathcal{A}$

$$A' \xrightarrow[\psi]{} A \xrightarrow[\phi]{} A''$$

This means $\ker \phi = \operatorname{Im} \psi$; and to say that

$$0 \longrightarrow A' \xrightarrow[\psi]{} A \xrightarrow[\phi]{} A'' \longrightarrow 0 \tag{1}$$

is exact means that ψ is injective, ϕ is surjective, and $\ker \phi = \operatorname{Im} \psi$. An exact sequence as in (1) is called a *short exact sequence.*

Suppose we have a family of functors $T^0, T^1, T^2, \ldots$, from $\mathcal{A}$ to the category of abelian groups. This family of functors is called a *δ-functor* (or *exact, connected sequence of functors*) iff

(1) For all short exact sequences as above and for all $n \geqq 0$, there exists a homomorphism $\delta_n : T^n(A'') \to T^{n+1}(A')$ so that the long sequence obtained by using the δ_n, viz:

$$0 \to T^0(A') \to T^0(A) \to T^0(A'') \xrightarrow{\delta_0} T^1(A') \to T^1(A) \to T^1(A'') \to$$

$$T^2(A') \to T^2(A) \to \cdots$$

is *exact* (in particular, we assume T^0 is left-exact!), and

(2) The big diagram

$$\begin{array}{ccccccccccccc} 0 \to & T^0(A') & \to & T^0(A) & \to \cdots \to & T^n(A) & \to & T^n(A'') & \to & T^{n+1}(A') & \to \cdots \\ & \downarrow & & \downarrow & & \downarrow & & \downarrow & & \downarrow & \\ 0 \to & T^0(B') & \to & T^0(B) & \to \cdots \to & T^n(B) & \to & T^n(B'') & \to & T^{n+1}(B') & \to \cdots \end{array}$$

(induced from the little commutative diagram

$$\begin{array}{ccccccccc} 0 \to & A' & \to & A & \to & A'' & \to 0 \\ & \downarrow & & \downarrow & & \downarrow & \\ 0 \to & B' & \to & B & \to & B'' & \to 0 \end{array} \quad)$$

is everywhere commutative.

We shall now construct the main δ-functor to be used in this course. Let G be a profinite group, let A be an abelian group on which G acts as a group of automorphisms. We thus have a map $G \times A \to A$ given by $\langle \sigma, a \rangle \to \sigma a$, where σa is the action of σ on a. If we give A the discrete topology, we shall say that A is a *G-module* if and only if the map $G \times A \to A$ is *continuous*.

Exercise: Given $a \in A$, let $U_a = \{\sigma \in G \mid \sigma a = a\}$ = the *stabilizer* of a in G. Then A is a G-module if and only if U_a is an *open* subgroup of G. If U is a given subgroup of G, let $A^U = \{a \in A \mid (\forall \sigma \in U)(\sigma a = a)\}$. Then, A is a G-module if and only if $A = \bigcup \{A^U \mid U \text{ is open in } G\}$.

Let $\mathcal{A}(G)$ be the category of G-modules and G-homomorphisms—it is an abelian category. Given an integer $n \geqq 0$, given $A \in \mathrm{Ob}\,\mathcal{A}(G)$, set

(2) $$C^n(G, A) = \{f \mid f\colon G^n \to A \text{ and } f \text{ is } \textit{continuous}\}.$$

Here, G^n means the cartesian product of G with itself n times and by G^0 we understand a set with one element $\{\emptyset\}$. Consequently, $C^0(G, A) \xrightarrow{\sim} A$.

Under pointwise operations $C^n(G, A)$ forms an abelian group, and, obviously, $C^n(G, -)$ is a functor from $\mathfrak{A}(G)$ to abelian groups.

Proposition 2. *The functor* $C^n(G,-)$ *is exact for all* n.

Proof: If $0 \to A' \to A \to A'' \to 0$ is exact, one sees trivially that

$$0 \to C^n(G, A') \to C^n(G, A) \xrightarrow{\phi} C^n(G, A'')$$

is exact. Given $f \in C^n(G, A'')$, consider any section of the homomorphism $A \to A''$, i.e., any map $\theta : A'' \to A$ which when composed with $A \to A''$ yields id. As A, A'' are discrete, θ is continuous; consequently, $\theta \circ f$ is continuous, hence, lies in $C^n(G, A)$. Clearly f is the image of $\theta \circ f$ under ϕ. Q.E.D.

If $f \in C^n(G, A)$, we define $\delta f \in C^{n+1}(G, A)$ by the formula

$$(3) \qquad \begin{aligned}(\delta f)(\sigma_1, \ldots, \sigma_{n+1}) &= \sigma_1 f(\sigma_2, \ldots, \sigma_{n+1}) \\ &+ \sum_{i=1}^{n} (-1)^i f(\sigma_1, \ldots, \sigma_i\sigma_{i+1}, \ldots, \sigma_{n+1}) \\ &+ (-1)^{n+1} f(\sigma_1, \ldots, \sigma_n) \ .\end{aligned}$$

Exercises: (a) $\delta : C^n \to C^{n+1}$ is a homomorphism for each A, in fact, it is a morphism of functors;

(b) $\delta\delta = 0$ i.e.,

$$(4) \qquad C^0(G, A) \overset{\delta}{\to} C^1(G, A) \overset{\delta}{\to} C^2(G, A) \overset{\delta}{\to} \ldots$$

is a *complex* (composition of two successive maps is zero).

The complex (4) is called the *standard complex*. Using the standard complex, we shall define our δ-functor. We set

(5)
$$\begin{aligned}
Z^n(G, A) &= \ker\,(C^n(G, A) \xrightarrow{\delta} C^{n+1}(G, A)) \\
&= \textit{group of } n \textit{ cocycles of } G \textit{ with coefficients in } A \\
B^n(G, A) &= \operatorname{Im}\,(C^{n-1}(G, A) \xrightarrow{\delta} C^n(G, A)) \\
&= \textit{group of } n \textit{ coboundaries of } G \textit{ with coeffs. in } A \\
H^n(G, A) &= Z^n(G, A)/B^n(G, A) \\
&= n^{th} \textit{ cohomology group of } G \textit{ with coefficients in } A.
\end{aligned}$$

PROPOSITION 3. *The sequence of functors* $\{H^n(G, -)\}$ *is a* δ*-functor on* $\mathfrak{A}(G)$ *to the category of abelian groups.*

Proof: It is completely trivial that a short exact sequence

$$0 \longrightarrow A' \longrightarrow A \longrightarrow A'' \longrightarrow 0$$

induces the exact sequences

$$0 \longrightarrow H^0(G, A') \longrightarrow H^0(G, A) \longrightarrow H^0(G, A''),$$

$$H^n(G, A') \longrightarrow H^n(G, A) \longrightarrow H^n(G, A''), \qquad n > 0,$$

and that a small diagram involving two short exact sequences yields commutative diagrams for all $n \geq 0$. To define the *connecting homomorphisms*, δ_n, we use the snake lemma and Proposition 2 as follows:

$$\begin{array}{ccccccccc}
0 & \longrightarrow & C^n(G, A') & \longrightarrow & C^n(G, A) & \longrightarrow & C^n(G, A'') & \longrightarrow & 0 \\
 & & \downarrow{\scriptstyle \delta_{A'}} & & \downarrow{\scriptstyle \delta_{A}} & & \downarrow{\scriptstyle \delta_{A''}} & & \\
0 & \longrightarrow & C^{n+1}(G, A') & \longrightarrow & C^{n+1}(G, A) & \xrightarrow{\pi} & C^{n+1}(G, A'') & \longrightarrow & 0
\end{array}$$

If $\bar{\xi} \in H^n(G, A'')$, choose $\xi \in Z^n(G, A'')$ representing it. There exists ξ_1 in $C^n(G, A)$ mapping onto ξ, and $\pi\delta_A(\xi_1) = \delta_{A''}(\xi) = 0$, so $\delta_A(\xi_1) = \eta_1$ is the image of some element η from $C^{n+1}(G, A')$. Now the image of $\delta_{A'}(\eta)$ in $C^{n+2}(G, A)$ is exactly $\delta_A(\eta_1) = 0$; therefore, $\delta_{A'}(\eta) = 0$, i.e.,

$\eta \in Z^{n+1}(G, A')$. Let $\bar{\eta}$ be its cohomology class, and set $\delta_n(\bar{\xi}) = \bar{\eta}$. One checks easily that δ_n is well-defined, and the rest of the proof is routine diagram chasing which will be left to the reader. Q.E.D.

Examples. (6) $C^0(G, A) = A$, and $(\delta a)(\sigma) = \sigma a - a$ as one sees from Equation (3). Hence, $Z^0(G, A) = \{a \in A \mid (\forall \sigma \in G)(\sigma a = a)\} = A^G$ = set of fixed points of A under the action of G. As $B^0(G, A) = (0)$ by definition, we conclude

$$H^0(G, A) = A^G \tag{6}$$

(7) If $f \in C^1(G, A)$, then $(\delta f)(\sigma, \tau) = \sigma f(\tau) - f(\sigma\tau) + f(\sigma)$. Hence, $f \in Z^1(G, A)$ if and only if

$$f(\sigma\tau) = \sigma f(\tau) + f(\sigma) .$$

Such functions are called *crossed homomorphisms*. If G acts trivially on A, i.e., $\sigma a = a$ for all $a \in A$, all $\sigma \in G$, then a crossed homomorphism is an ordinary one. $B^1(G, A)$ consists of those $f(\sigma)$ having the form $f(\sigma) = \sigma a - a$ for some $a \in A$—these are the *principal crossed homomorphisms*. In the case of trivial action, a principal crossed homomorphism is identically zero, so that

$$\begin{aligned} H^1(G, A) &= \text{crossed homos/principal crossed homos} \\ H^1(G, A) &= \mathrm{Hom}_{\mathrm{cont}}(G, A) \text{ if } G \text{ acts trivially on } A . \end{aligned} \tag{7}$$

There can be many δ-functors on a category to abelian groups, but they count for little unless they have a uniqueness property. It is to this that we now turn.

Grothendieck [GT] introduced the notion of universal δ-functor for the purposes of uniqueness. We say a δ-functor $\{T^n\}$ is a *universal δ-functor* if and only if for *any* other δ-functor $\{S^n\}$ and *any* morphism $f_0\colon T^0 \to S^0$ of functors, *there exists a unique morphism* (of δ-functors): $\{T^n\} \to \{S^n\}$ *extending* f_0. Here, a morphism of δ-functors is, of course, a sequence of morphisms f_n: $T^n \to S^n$ which commute with the δ's of the families $\{T^n\}$ and $\{S^n\}$.

Now we have

THEOREM 5 (Uniqueness Theorem). *Let* $\{S^n\}$ *and* $\{T^n\}$ *be two universal* δ*-functors from* $\mathfrak{A}$ *to the category of abelian groups. If* S^0 *is isomorphic to* T^0*, then* $\{S^n\}$ *is isomorphic to* $\{T^n\}$ *by a uniquely determined isomorphism.*

Proof: We have inverse isomorphisms $f_0 : S^0 \to T^0$ and $g_0 : T^0 \to S^0$. As both are universal δ-functors, there exist uniquely determined morphisms $f: S \to T$ and $g: T \to S$ extending f_0 and g_0 respectively. It follows that fg and gf extend f_0g_0 and g_0f_0 respectively. However, f_0g_0 and g_0f_0 are the respective identities from T to T and S to S, and, as such, are extendible by the proper identity morphism. Thus, the uniqueness of extensions yields $fg = id_T$, $gf = id_S$. Q.E.D.

Clearly the problem of uniqueness of a δ-functor is now reduced to finding some reasonable condition for universality. The standard such criterion is due to Grothendieck [GT], and one form of it is

THEOREM 6. (Criterion of Effacability for uniqueness of δ-functors). *Let* $\{T^n\}$ *be a* δ*-functor and suppose for each* $n > 0$ *there is a functor* E_n *from* $\mathfrak{A}$ *to itself and a monomorphism of functors* $Id_{\mathfrak{A}} \to E_n$ *such that the induced morphism of functors* $T^n \to T^n \circ E_n$ *is the zero morphism. (In more concrete terms, we assume that for every object* A *of* $\mathfrak{A}$ *and every* $n > 0$*, there is a monomorphism of* A *to some object* $E_n(A)$ *functorially associated to* A*, say* $u_n : A \to E_n(A)$*, such that the induced map* $u_{n*} : T^n(A) \to T^n(E_n(A))$ *is zero.) Then the* δ*-functor* $\{T^n\}$ *is universal.*

Proof: Let $\{S^n\}$ be a given δ-functor and $f_0 : T^0 \to S^0$ a given morphism of functors. By induction we assume an extension of f_0 has been found up to and including $f_n : T^n \to S^n$. We shall show how to push the argument one step further. By hypothesis, given A, there exists $E_{n+1}(A)$ and a monomorphism $A \to E_{n+1}(A)$, all functorial in A. Consider the short exact sequence

$$0 \to A \xrightarrow{u_{n+1}} E_{n+1}(A) \longrightarrow C \longrightarrow 0$$

where $C = \operatorname{coker} u_{n+1}$. We obtain a commutative diagram

$$\begin{array}{ccccccc}
T^n(E_{n+1}(A)) & \longrightarrow & T^n(C) & \xrightarrow{\delta_n} & T^{n+1}(A) & \xrightarrow{u_{n+1*}} & T^{n+1}(E_{n+1}(A)) \\
\downarrow f_n & & \downarrow f_n & & \vdots & & \vdots \\
S^n(E_{n+1}(A)) & \longrightarrow & S^n(C) & \xrightarrow{\delta_n} & S^{n+1}(A) & \xrightarrow{u_{n+1*}} & S^{n+1}(E_{n+1}(A))
\end{array}$$

by induction hypothesis, and we must construct the first dotted arrow. Let $\xi \in T^{n+1}(A)$, then $u_{n+1*}(\xi) = 0$ so there exists $\eta \in T^n(C)$ with $\delta_n(\eta) = \xi$. Let $f_{n+1}(\xi) = \delta_n f_n(\eta)$. If $\eta' \in T^n(C)$ also yields ξ after application by δ_n, then $\eta - \eta'$ is killed by δ_n, so is the image of some ν in $T^n(E_{n+1}(A))$. But then, $\delta_n(f_n(\nu)) = 0$, so that $\delta_n f_n(\eta - \eta') = 0$; i.e., f_{n+1} is well-defined. Because E_{n+1} is functorial in A, one checks that f_{n+1} is a morphism of functors $T^{n+1} \to S^{n+1}$, and this shows that induction yields a unique extension of f_0 to all levels $n > 0$. Q.E.D.

Remarks. (1) Much stronger criteria of effacibility are available, see [GT] or [B]. In these, the functoriality of f_{n+1} is not at all obvious and must be proved by a sort of "common refinement" argument.

(2) The object $E_n(A)$ is called an *effacing object* and u_n is an *effacing monomorphism.* The origin of the terms is obvious in view of what the morphism u_{n*} does to ξ.

COROLLARY 1. *Let* $\{T^n\}$ *be a* δ*-functor and assume that for all* A *in* $\mathfrak{A}$, *there exists a monomorphism of* A *into some* T*-acyclic object* $\mathfrak{M}(A)$. *(An object* B *is* T*-acyclic if* $T^n(B) = (0)$ *for* $n > 0$.*) If* $\mathfrak{M}(A)$ *is a functor of* A, *then* $\{T^n\}$ *is universal.*

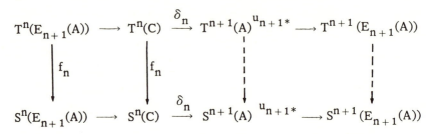

COROLLARY 2. *Under the hypotheses of* Cor. 1, *every injective object of* $\mathfrak{A}$ *is* T-*acyclic.*

Proof: Let Q be injective in $\mathfrak{A}$. According to Cor. 1, there is an exact sequence $0 \to Q \to \mathfrak{M} \to C \to 0$ where $\mathfrak{M}$ is T-acyclic. This sequence splits because Q is injective, so $\mathfrak{M} = Q \oplus C$. Thus

$$(0) = T^n(\mathfrak{M}) = T^n(Q) \oplus T^n(C) \quad \text{for} \quad n > 0,$$

and so $T^n(Q) = (0)$ for $n > 0$. Q.E.D.

In the next section, we shall show that the δ-functor $\{H^n(G, -)\}$ constructed above is universal.

§2. Behavior of the cohomology as a functor, induced modules.

Suppose G, G′ are profinite groups and A, A′ are G, resp. G′, modules. Let us assume given a pair of group homomorphisms $\phi : G' \to G$ (continuous) and $\psi : A \to A'$. We shall say $<\phi, \psi>$ is a *compatible pair* if and only if $\sigma'(\psi(a)) = \psi(\phi(\sigma')(a))$ for every $a \in A$ and $\sigma' \in G'$. That is, if and only if the diagram

$$\begin{array}{ccc} G \times A & \longrightarrow & A \\ \phi \uparrow \; \downarrow \psi & & \downarrow \psi \\ G' \times A' & \longrightarrow & A' \end{array}$$

commutes in the obvious sense. If $f \in C^n(G, A)$, then $\psi \circ f \circ \phi$ belongs to $C^n(G', A')$, and $f \to \psi \circ f \circ \phi$ is a *cochain map,* meaning that it commutes with δ. This being the case, one sees that we get a morphism of δ-functors $H^*(\phi, \psi) : H^*(G, A) \to H^*(G', A')$ from any compatible pair $<\phi, \psi>$. (Here, as is usual, the upper asterisk notation signifies simultaneous maps $H^n(G, A) \to H^n(G', A')$ for each $n \geqq 0$.)

There are two important compatible pairs used with enormous frequency in the subject, they are

(1) Let S be a subgroup of G, let A be a G-module. Then A is an S-module, and the inclusion $S \to G$ is compatible with the identity $A \to A$. We obtain a homomorphism

$$\text{res:}\quad H^*(G, A) \longrightarrow H^*(S, A)$$

called *restriction.* (The name comes from the obvious action of restricting a cocycle from G to S.)

(2) Let S be a closed *normal* subgroup of G, and let $\phi: G \to G/S$ be the natural projection. If A is a G-module, then A^S is a G/S -module in a natural way, and the inclusion map $\psi: A^S \hookrightarrow A$ is compatible with ϕ. We obtain the homomorphism

$$\text{inf}: H^*(G/S, A^S) \longrightarrow H^*(G, A)$$

called *inflation.* Explicitly, if $f(\bar{\sigma}_1, \ldots, \bar{\sigma}_n)$ is a G/S-cocycle in A^S, then if we set $f(\sigma_1, \ldots, \sigma_n) = f(\bar{\sigma}_1, \ldots, \bar{\sigma}_n)$ considered in A, we obtain the inflation.

THEOREM 7. *Let* $\{G_\alpha\}$ *be a projective mapping family of profinite groups, let* $\{A_\alpha\}$ *be a direct mapping family of* G_α*-modules, and assume that for all* $\beta \geqq \alpha$, *the maps* $\phi_\beta^\alpha: G_\beta \to G_\alpha$, $\psi_\alpha^\beta: A_\alpha \to A_\beta$ *are compatible. Let* $G = \text{proj}\lim_\alpha G_\alpha$, $A = \text{dir}\lim_\alpha A_\alpha$, *then there is a unique, continuous operation of* G *on* A *so that for every* α, *the maps* $\phi_\alpha: G \to G_\alpha$, $\psi_\alpha: A_\alpha \to A$ *are compatible—hence,* A *is a* G*-module. We have the following formula for the cohomology of* G *with coefficients in* A

$$H^*(G, A) = H^*(\text{proj}\lim_\alpha G_\alpha, \text{dir}\lim_\alpha A_\alpha) = \text{dir}\lim_\alpha H^*(G_\alpha, A_\alpha)\,.$$

Proof: Observe first that as $a \in A$ implies that there exists an α with $a = \psi_\alpha(a_\alpha)$ (some $a_\alpha \in A_\alpha$), the uniqueness of the G-module structure on A is forced by the compatibility requirement for $\langle\phi_\alpha, \psi_\alpha\rangle$. Given $\sigma = \langle\sigma_\alpha\rangle_\alpha \in G$ and $a \in A$ [say $a = \psi_\beta(a_\beta)$ for some $a_\beta \in A_\beta$], consider

$\sigma_\beta a_\beta \in A_\beta$, and set $\sigma a = \psi_\beta(\sigma_\beta a_\beta)$. If $\gamma \geqq \beta$, then $\phi_\gamma{}^\beta(\sigma_\gamma) = \sigma_\beta$; hence,

$$\begin{aligned}\psi_\beta(\sigma_\beta a_\beta) &= \psi_\beta((\phi_\gamma{}^\beta \sigma_\gamma) a_\beta) = \psi_\gamma \psi_\beta{}^\gamma((\phi_\gamma{}^\beta \sigma_\gamma) a_\beta) \\ &= \psi_\gamma(\sigma_\gamma(\psi_\beta{}^\gamma a_\beta)) = \psi_\gamma(\sigma_\gamma a_\gamma) \ ,\end{aligned}$$

where $\psi_\beta{}^\gamma a_\beta = a_\gamma$, hence $\psi_\gamma(a_\gamma) = a$, as well. It follows from these equations that the definition of σa is unambiguous, and the fact that A is now a G-module is routine checking.

For the statement about cohomology, all one need prove is that $C^n(G, A) = \operatorname{dir\,lim}_\alpha C^n(G_\alpha, A_\alpha)$, for all n. If this is proved, the fact that "cohomology commutes with direct limits" and the compatibility of the various maps will yield our theorem instantly. Since $<\phi_\alpha, \psi_\alpha>$ is a compatible pair, it yields a map

$$i_\alpha : C^n(G_\alpha, A_\alpha) \longrightarrow C^n(G, A) \ .$$

If Γ denotes $\operatorname{dir\,lim}_\alpha C^n(G_\alpha, A_\alpha)$, then the DUMP implies there exists a map f making the diagram

$$\begin{array}{ccc} \Gamma & \xrightarrow{\ f\ } & C^n(G, A) \\ {\scriptstyle c_\alpha}\nwarrow & & \nearrow{\scriptstyle i_\alpha} \\ & C^n(G_\alpha, A_\alpha) & \end{array}$$

commute for all α —where c_α denotes the universal map at the αth stage. If $g \in \Gamma$ is mapped to zero by f, then there exists an index α, and an element $g_\alpha \in C^n(G_\alpha, A_\alpha)$ such that $i_\alpha(g_\alpha) = 0$. But, this means that

$$0 = i_\alpha g_\alpha(\sigma_1, \ldots, \sigma_n) = \psi_\alpha(g_\alpha(\sigma_{1\alpha}, \ldots, \sigma_{n\alpha})) \ .$$

However, as G_α is compact and A_α is discrete only *finitely* many elements of A_α occur as $g_\alpha(\sigma_{1\alpha}, \ldots, \sigma_{n\alpha})$ no matter what the choice of $\sigma_1, \ldots, \sigma_n$. By choosing β large enough, we may clearly arrange to have

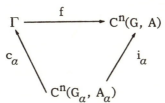

$g_\beta(\sigma_{1\beta}, \ldots, \sigma_{n\beta}) = 0$ for all $\sigma_{1\beta}, \ldots, \sigma_{n\beta}$ which come from projecting arbitrary elements of G down to β. The same holds for all indices larger than β. From this, one sees easily that g_β goes to zero in Γ, i.e., f is injective. For the surjectivity, let $h \in C^n(G, A)$ be given. There are only *finitely* many values of h in A, so clearly there exists an index α for which $i_\alpha(h_\alpha) = h$ for some $h_\alpha \in C^n(G_\alpha, A_\alpha)$. The image of h_α in Γ is the required pre-image for h in Γ. Q.E.D.

Observe how the continuity of our cochains came into play in the above proof. The theorem is *false* for discontinuous cochains.

COROLLARY 1. *If* G *is a profinite group and* A *is a* G-*module, then*

$$H^*(G, A) = \text{dir lim}\,\{H^*(G/U, A^U) \mid U \text{ open, normal in } G\}.$$

The maps in the family $\{H^*(G/U, A^U)\}$ *are inflation.*

COROLLARY 2. *If* G *is a profinite group,* S *a closed subgroup of* G, *and* $N \lhd S$, *then*

$$H^*(S/N, A^N) = \operatorname*{dir\,lim}_\alpha H^*(S_\alpha/N_\alpha, A^{N_\alpha})$$

where $S_\alpha = U_\alpha \cap S$, $N_\alpha = U_\alpha \cap N$, *and* U_α *ranges over all the open, normal subgroups of* G.

COROLLARY 3. *Let* S *be a closed subgroup of the profinite group* G, *and let* $\{S_\alpha\}$ *be the set of open subgroups of* G *containing* S. *If* A *is a* G-*module, then*

$$H^*(S, A) = \operatorname*{dir\,lim}_\alpha H^*(S_\alpha, A)$$

where the maps in the family $\{H^*(S_\alpha, A)\}$ *are restriction.*

COROLLARY 4. *If* G *is a profinite group and* A *is a* G-*module, let* $\{A_\alpha\}$ *denote the family of all submodules of* A *which are finitely generated as* G-*modules. Then*

$$H^*(G, A) = \operatorname*{dir\,lim}_\alpha H^n(G, A_\alpha)\ .$$

Exercise. A G-module A is finitely generated as G-module if and only if it is finitely generated as abelian group.

Remark. Corollary 1 gives the connection between the cohomology theory of finite groups (see [CE, LR, SCL]) and that for profinite groups.

Let G be a profinite group and let S be a closed subgroup of G. There is a functor from $\mathfrak{A}(G)$ to $\mathfrak{A}(S)$, denoted $\pi^*_{S\to G}$ (or simply π^* if no confusion results), namely $\pi^*_{S\to G}(B) = B$ regarded as S-module. Let us call $\pi^*(B)$ the *inverse image* of B. The key to proving the universality of our cohomology is a functor in the other direction i.e., from $\mathfrak{A}(S)$ to $\mathfrak{A}(G)$. If B is an S-module, let $\pi_{*S\to G}(B)$ be the group (under pointwise addition) of all continuous functions from G to B such that $f(\sigma\tau) = \sigma f(\tau)$ if $\sigma \in S$. Thus,

$$\pi_{*S\to G}(B) = \{f: G \to B \mid f \text{ is cont. and } f(\sigma\tau) = \sigma f(\tau) \text{ if } \sigma \in S\} .$$

The group $\pi_{*S\to G}(B)$ will be made into a G-module, and then we shall call it the *induced module* from S to G of B. In Tate's original paper on the subject (see [D]), he denoted $\pi_{*S\to G}(B)$ by $\mathrm{Map}_S(G, B)$. In keeping with standard terminology, we will also call $\pi_{*S\to G}(B)$ the *direct image* of B.

PROPOSITION 4. *Let* G *be a profinite group, let* S *be a closed subgroup of* G *and let* B *be an* S*-module. The operation*

$$(\rho f)(\tau) = f(\tau\rho)$$

makes $\pi_{*S\to G}(B)$ *a* G*-module, and* $\pi_{*S\to G}$ *is a functor from* $\mathfrak{A}(S)$ *to* $\mathfrak{A}(G)$. *It satisfies the following universal mapping property: For all* G*-modules* A *and* S*-modules* B,

$$\mathrm{Hom}_G(A, \pi_{*S\to G}(B)) \xrightarrow{\sim} \mathrm{Hom}_S(\pi^*_{S\to G}(A), B)$$

In other words, π_* *and* π^* *are adjoint functors.*

Proof: That $(\rho f)(\tau) = f(\tau\rho)$ is a G-module action is trivial, except for the continuity. For this, we must show that

$$U_f = \{\rho \in G \mid \rho f = f\}$$

is open in G. Now f is continuous, so it has only finitely many values in B, say $b_1, \ldots, b_r$. If U_j is the inverse image (under f) of b_j in G, and if χ_j is the function on G which is identically zero off U_j and identically b_j on U_j, then χ_j is continuous on G and

$$f = \Sigma_j \chi_j \ .$$

Let $\mathcal{O}_j = \{\rho \in G \mid \rho\chi_j = \chi_j\}$, and let $\mathcal{O} = \bigcap_j \mathcal{O}_j$. Clearly, $\mathcal{O} \subseteq U_f$, and as U_f is closed, if we prove each $\mathcal{O}_j$ is open it will follow that $\mathcal{O}$ is open; hence, that U_f is open. Therefore, we may and do assume that $r = 2$, i.e., that f has one non-zero value on G. Let V be the set where f is non-zero, then its complement V^c is the set where f vanishes. Both V and V^c are open and closed, hence, compact and open subsets of G by the continuity of f. We must show that $\{\rho \in G \mid V\rho \subseteq V,\ V^c\rho \subseteq V\}$ is open. But this is a standard consequence of the compactness and openness of V.

If $g \in \mathrm{Hom}_G(A, \pi*_{S \to G}(B))$, then $g(a)(\tau) \in B$, and if $\tau \in S$ we have

$$g(a)(\tau) = \tau \cdot g(a)(1) \ .$$

Set $(\theta g)(a) = g(a)(1)$, then one checks easily that θ is the required map from $\mathrm{Hom}_G(A, \pi*_{S \to G}(B))$ to $\mathrm{Hom}_S(\pi^*_{S \to G}(A), B)$. To prove that θ is an isomorphism, define the map ϕ in the opposite direction from θ *via* the equation

$$[(\phi h)(a)](\tau) = h(\tau a) \ ,$$

here $h \in \mathrm{Hom}_S(\pi^*_{S \to G} A, B)$, and $\tau \in G$. Trivial computations show that ϕ and θ are inverse to each other. Q.E.D.

COROLLARY 1. *If* B *is an injective* **Z**-*module* (= *divisible abelian group*), *then* $\pi*_{1 \to G}(B)$ *is an injective* G-*module.*

Proof: By hypothesis, the functor $A \rightsquigarrow \mathrm{Hom}_Z(\pi^*_{1 \to G}(A), B)$ is exact (since π^* is an exact functor). However, by our proposition above this means that $A \rightsquigarrow \mathrm{Hom}_G(A, \pi_{*1\to G}(B))$ is an exact functor. We conclude that the module $\pi_{*1\to G}(B)$ is injective.

COROLLARY 2. *If* A *is a* G*-module, the map*

$$\varepsilon_A : A \longrightarrow \pi_{*1\to G}(\pi^*_{1\to G}(A))$$

given by $\varepsilon_A(a)(\tau) = \tau a$, *is a monomorphism of* G*-modules. Consequently, the category* $\mathfrak{A}(G)$ *possesses enough injectives, i.e., every* G*-module may be embedded in an injective* G*-module.*

Proof: The first assertion is a simple, explicit computation. As for the injectives, if A is a given G-module, then $\pi^*_{1\to G}(A)$ may be embedded in an injective abelian group, as is well-known. Call this group Q. The exact sequence $0 \to \pi^*(A) \to Q$ yields the exact sequence

$$0 \longrightarrow \pi_*\pi^*(A) \xrightarrow{\ \phi\ } \pi_*(Q)$$

of G-modules. Now composing $\varepsilon_A : A \to \pi_*\pi^*(A)$ and ϕ we obtain the embedding $A \to \pi_*(Q)$, which completes the proof according to Cor. 1. Q.E.D.

We are now very close to the universality of the chosen δ-functor $\{H^n(G,-)\}$. First we shall show this universality is equivalent to several other statements, and this will give us the technical freedom necessary to finally prove universality.

PROPOSITION 5. *The following four statements are all equivalent for every profinite group* G.

(1) *The* δ*-functor* $\{H^n(G,-)\}$ *on* $\mathfrak{A}(G)$ *to abelian groups is universal for all* G ;

(2) $H^n(G,-)$ *is the* n^{th} *right derived functor of the left exact functor* $H^0(G,-) = \Gamma, (\Gamma(A) = A^G)$ *for all* G;

(3) *For every* G, *every closed subgroup* S *of* G, *and every* S*-module* B, *the map*

$$\mathrm{sh}: H^n(G, \pi_{*S\to G}(B)) \longrightarrow H^n(S, B)$$

induced by the compatible maps $S \hookrightarrow G$ *and* $\pi_*(B) \to B$ *(via* $f \mapsto f(1)$*) is an isomorphism for every* n.

(4) *For every* G, *every abelian group* B, *and every positive integer* r, *the group* $H^r(G, \pi_{*1 \to G}(B))$ *is trivial.*

Proof: (1) $\Longrightarrow$ (2) Since $\mathcal{A}(G)$ has enough injectives, $\{R^n\Gamma\}_{n=0}^{\infty}$ is a universal δ-functor (cf. [CE]). But so is $\{H^n(G,-)\}$, and the two agree in dimension zero; hence, the uniqueness theorem implies (2).

(2) $\Longrightarrow$ (3) By hypothesis, $H^n = R^n\Gamma$ for each G; hence, as $\{R^n\Gamma\}$ is a universal δ-functor, so is $\{H^n(G,-)\}$ for each G. Thus $\{H^n(S,-)\}$ is a universal δ-functor on $\mathcal{A}(S)$. But, we claim that the δ-functor $\{H^n(G, \pi_{*S \to G}(-))\}$ on $\mathcal{A}(S)$ to abelian groups is also universal. (It is a δ-functor because one can easily show π_* is an exact functor. This fact alone is of tremendous importance for (3) as we shall see in §4.) Given $B \in \mathrm{Ob}\,\mathcal{A}(S)$, there is an injective S-module J, and a monomorphism $B \to J$. Hence, $0 \to \pi_{*S \to G}(B) \to \pi_{*S \to G}(J)$ is also exact, and $\pi_* J$ is injective in $\mathcal{A}(G)$. Now by (2), as $R^n\Gamma$ vanishes on injectives for $n > 0$, we deduce that $H^n(G, \pi_* J) = (0)$ for $n > 0$. It follows that $\{H^n(G, \pi_*(-))\}$ is effacable, and Th'm 6 now shows that it is universal. Conclusion (3) will follow when we prove that the two functors agree in dimension zero under the map described above. This is a very simple calculation and will be left to the reader.

(3) $\Longrightarrow$ (4) By (3), for every abelian group B,

$$H^r(G, \pi_{*1 \to G}(B)) \xrightarrow{\sim} H^r(\{1\}, B).$$

However, one checks, directly from the definition of H^r, that $H^r(\{1\}, B)$ vanishes for $r > 0$. (Observe that $C^0 = C^1 = C^2 = \cdots = C^n = \cdots$ for all n.)

(4) $\Longrightarrow$ (1). By Cor. 2 above, every G-module A admits an injection

$$0 \to A \to \pi_{*1 \to G}(\pi^*_{1 \to G}(A)) .$$

If we apply (4) to the abelian group $B = \pi^*_{1\to G}(A)$, we deduce that $\{H^n(G,-)\}$ is an effacable δ-functor; hence, Theorem 6 implies its universality. Q.E.D.

THEOREM 8 (Shapiro's Lemma). *For any profinite group* G, *any closed subgroup* S *of* G, *and any* S-*module* B, *the Shapiro homomorphism,* sh, *of* (3) *above*

$$\text{sh}: H^n(G, \pi_{*S\to G}(B)) \longrightarrow H^n(S,B)$$

is an isomorphism for all $n \geqq 0$. *Consequently, all of* (1), (2), (3), (4) *hold, and the cohomology groups form the unique, universal* δ*-functor from* $\mathcal{A}(G)$ *to abelian groups having* $H^0(G,-) = -^G$.

Proof: Actually, we shall prove (4), then Proposition 5 above yields everything else. Let $f(\sigma_1,\ldots,\sigma_n)$ be an n-cocycle $(n>0)$ for G with coefficients in $\pi_{*1\to G}(B)$. Define $g(\sigma_1,\ldots,\sigma_{n-1}) \in \pi_{*1\to G}(B)$ by the formula

$$g(\sigma_1,\ldots,\sigma_{n-1})(\sigma) = f(\sigma,\sigma_1,\ldots,\sigma_{n-1})(1) \ .$$

We claim $\delta g = f$; this will complete the proof. Now

$$\delta g(\sigma_1,\ldots,\sigma_n) = \sigma_1 g(\sigma_2,\ldots,\sigma_n) + \sum_{j=1}^{n-1} (-1)^j\, g(\sigma_1,\ldots,\sigma_j\sigma_{j+1},\ldots,\sigma_n)$$
$$+ (-1)^n g(\sigma_1,\ldots,\sigma_{n-1}) \ .$$

Hence,

$$\delta g(\sigma_1,\ldots,\sigma_n)(\tau) = g(\sigma_2,\ldots,\sigma_n)(\tau\sigma_1) + \sum_{j=1}^{n-1} (-1)^j g(\sigma_1,\ldots,\sigma_j\,\sigma_{j+1},\ldots,\sigma_n)(\tau)$$
$$+ (-1)^n g(\sigma_1,\ldots,\sigma_{n-1})(\tau)$$

$$= f(\tau\sigma_1,\sigma_2,\ldots,\sigma_n)(1) + \sum_{j=1}^{n-1} (-1)^j f(\tau,\sigma_1,\ldots,\sigma_j\sigma_{j+1},\ldots,\sigma_n)(1)$$
$$+ (-1)^n f(\tau,\sigma_1,\ldots,\sigma_{n-1})(1) \ .$$

Now, $\delta f(\tau, \sigma_1, \ldots, \sigma_n)(1) = 0$ implies

$$0 = (\tau f)(\sigma_1, \ldots, \sigma_n)(1) - f(\tau\sigma_1, \sigma_2, \ldots, \sigma_n)(1)$$
$$+ \sum_{j=1}^{n-1} (-1)^{j+1} f(\tau, \sigma_1, \sigma_2, \ldots, \sigma_j\sigma_{j+1}, \ldots, \sigma_n)(1)$$
$$+ (-1)^{n+1} f(\tau, \sigma_1, \ldots, \sigma_{n-1})(1) \ .$$

Consequently,

$$\delta g(\sigma_1, \ldots, \sigma_n)(\tau) = (\tau f)(\sigma_1, \ldots, \sigma_n)(1) = f(\sigma_1, \ldots, \sigma_n)(\tau) \ .$$

Q.E.D.

Besides the application to universality of the cohomology, Shapiro's lemma serves very effectively in a process known as *décallage* or *dimension-shifting*. This is a process whereby results true for the cohomology groups of all modules in a certain dimension are proved true for all higher dimensions. As a first example, we prove

PROPOSITION 6. *Let* G *be a profinite group. A necessary and sufficient condition that*

$$H^n(G, B) = (0) \text{ for all } n > r, \text{ and all } G\text{-modules } B$$

is that $H^{r+1}(G, B) = (0)$, *for all* B. *The same conclusion holds if we restrict attention to the torsion* G*-modules.*

Proof: We shall show that given n, if $H^n(G, B)$ vanishes for all B, then $H^{n+1}(G, B)$ also vanishes for all B. This will then complete the non-torsion part of the proof. Given any B, consider the exact sequence of G-modules

$$0 \longrightarrow B \xrightarrow{\varepsilon_B} \pi_*\pi^*(B) \longrightarrow \text{coker } \varepsilon_B \longrightarrow 0 \ ,$$

and apply cohomology.

$$H^n(G, \pi_*\pi^*B) \to H^n(G, \text{coker } \varepsilon_B) \xrightarrow{\delta} H^{n+1}(G, B) \to H^{n+1}(G, \pi_*\pi^*B)$$

Shapiro's lemma shows the extremes of our sequence vanish for every B; hence, the result for n implies the result for n + 1, as contended.

As for torsion modules, all we need check is that $\pi_{*1\to G}(B)$ is a torsion module if B is a torsion group. However, if $f \in \pi_*(B)$, as f is continuous it has only finitely many values $b_1, \ldots, b_n$. Each b_j is a torsion element of B, say of exponent r_j. Clearly $rf = 0$, where r = product of the r_j. Q.E.D.

Other examples of this powerful technique will occur later, and in profusion. They, more than any general discussion, will explain the technique and its uses.

§3. Restriction, transfer (co-restriction), and cup-products.

In this section, we shall exploit in a very heavy manner the universality of the δ-functor $\{H^n(G, -)\}$. Our first task is to rework the definition of the restriction maps to show how they fit into this framework.

PROPOSITION 7. *Let* G *be a profinite group, let* S *be a closed subgroup, and let* A *be a* G*-module. For every* $n \geqq 0$, *the composition*

$$H^n(G, A) \to H^n(G, \pi_{*S\to G}(\pi^*_{S\to G}(A))) \longrightarrow H^n(S, \pi^*_{S\to G}(A))$$

coincides with restriction.

Proof: Call the composition, res*. Then a trivial computation shows that for $n = 0$, res* = res. However, both are maps of a universal δ-functor to another δ-functor; hence, they agree for all n. Q.E.D.

PROPOSITION 8. *Let* G *be a profinite group, let* S *be an open subgroup, and let* A *be a* G*-module. Then for every* $n \geqq 0$, *there is a unique homomorphism*

$$\text{tr}: H^n(S, \pi^*_{S\to G}(A)) \longrightarrow H^n(G, A)$$

called the transfer or co-restriction, so that in dimension zero tr: $A^S \to A^G$ *is given by* $a \mapsto \Sigma_\rho\, \rho a$ *where* ρ *ranges over a transversal for* S *in* G. *Moreover,* tr *agrees with the map given by the composition*

$$H^n(S, \pi^*_{S\to G}(A)) \xrightarrow{\sim} H^n(G, \pi_{*S\to G}(\pi^*_{S\to G}(A))) \longrightarrow H^n(G, A)$$

which is induced by the G*-homomorphism* $\pi_{*S\to G}(\pi^*_{S\to G}(A)) \to A$ *given by* $f \mapsto \Sigma_\rho\, \rho \cdot (f(\rho^{-1}))$, *and* ρ *is as above.*

Proof: The map $a \mapsto \Sigma\, \rho a$ taking $A^S \to A^G$ is well-defined, for if $\{\rho'\}$ is another coset representative family, then $\rho = \rho'\sigma$ for some $\sigma \in S$. Hence, $\rho a = \rho'\sigma a = \rho' a$, and our claim follows. By universality, the maps tr exist as claimed. Moreover, if we show that $f \mapsto \Sigma_\rho\, \rho \cdot (f(\rho^{-1}))$ is well-defined, then in dimension zero we recover the map $a \mapsto \Sigma\, \rho a$, and so all will follow by universality once again. (Observe that if A is an injective G-module, then $\pi^*_{S\to G}\, A$ is an injective S-module as $(G:S) < \infty$; hence, the δ-functor $\{H^n(S, \pi^*(-))\}$ is indeed universal.) Let $\{\rho'\}$ be another transversal as before, then we can write $\rho = \rho'\sigma$ for $\sigma \in S$. Hence $\rho \cdot f(\rho^{-1}) = \rho'\sigma \cdot f(\sigma^{-1}\rho'^{-1})$. As $f \in \pi_{*S\to G}A$, $f(\sigma^{-1}\rho'^{-1}) = \sigma^{-1}f(\rho'^{-1})$; therefore, $\rho \cdot f(\rho^{-1}) = \rho' \cdot f(\rho'^{-1})$, and we are done. Q.E.D.

If we put together Propositions 7 and 8, we get the following extremely useful proposition.

PROPOSITION 9. *Let* G *be a profinite group and let* S *be an open subgroup of* G. *For every* G*-module* A, *the composed mapping*

$$\mathrm{tr} \circ \mathrm{res}:\ H^r(G, A) \longrightarrow H^r(S, \pi^* A) \longrightarrow H^r(G, A)$$

is equal to multiplication by the index $(G:S)$, *for every* $r \geqq 0$. *Succinctly,*

$$\mathrm{tr} \circ \mathrm{res} = \mathrm{index}.$$

Proof: In dimension zero, $\mathrm{tr} \circ \mathrm{res}$ is given by the composition of the inclusion $A^G \to A^S$ and the "norm" $\mathfrak{N}: A^S \to A^G$. That is, for $r = 0$,

$$(\mathrm{tr} \circ \mathrm{res})(a) = \Sigma_\rho\, \rho a = (G:S)a\ .$$

Thus, multiplication by (G : S) and tr ∘ res are two maps of the universal δ-functor which agree in dimension zero. By the uniqueness theorem, our proposition is now proved. Q.E.D.

COROLLARY 1. *Let* G *be a finite group whose order is* g, *and let* A *be an arbitrary (not necessarily torsion)* G-*module. Then for every* $r > 0$,

$$g \cdot H^r(G, A) = (0) \ .$$

Proof: Apply Proposition 9 to the case $S = \{1\}$. Then $g = (G : S)$, and $H^r(S, \pi^* A) = (0)$, for $r > 0$. Q.E.D.

COROLLARY 2. *Let* G *be a profinite group and let* A *be any* G-*module. Then* $H^r(G, A)$ *is a torsion group for all positive* r.

According to Corollary 2, we can write

$$H^r(G, A) = \coprod_p H^r(G, A; p), \qquad r > 0$$

where $\coprod$ means direct sum, p ranges over all prime numbers, and $H^r(G, A; p)$ means the p-primary component of $H^r(G, A)$. This has the following consequence:

PROPOSITION 10. *Let* G *be a profinite group, let* S *be a closed subgroup, and let* A *be a* G-*module. Suppose that* (G : S) *is prime-to-*p *for the prime number* p. *Then for every integer* $r \geqq 0$,

$$\text{res} : H^r(G, A; p) \longrightarrow H^r(S, \pi^* A; p)$$

is an injection. In particular, if $S = G_p$ *—a* p-*Sylow subgroup of* G, *then*

$$\text{res} : H^r(G, A; p) \longrightarrow H^r(G_p, \pi^* A; p)$$

is injective for all $r \geqq 0$. *Consequently, if* G *is a* p-*group, then* $H^r(G, A)$ *is a* p-*torsion group for all positive* r, *and every* G-*module* A.

Proof: We know (Corollary 3 of Theorem 7) that

$$H^r(S, \pi^* A ; p) = \varinjlim_{\text{restrictions}} H^r(U, \pi^*_{U \to G} A ; p)$$

where U ranges over all *open* subgroups containing S. It follows immediately that we may assume S is an open subgroup of G. In this case, tr is defined, and we have for any ξ killed by restriction:

$$0 = \operatorname{tr}(\operatorname{res}(\xi)) = (G : S)(\xi) \ .$$

As $(G:S)$ is prime-to-p, and as ξ is a p-torsion element, we deduce $\xi = 0$. If G is a p-group, then for every prime number $q \neq p$, we have $G_q = \{1\}$. Hence,

$$\operatorname{res} : H^r(G, A ; q) \longrightarrow H^r(\{1\}, A ; q) = (0)$$

is an injection for $r > 0$. It follows that $H^r(G, A)$ is a p-group for positive r. Q.E.D.

Using example (7) and the above, we can make several observations:

If A is a torsion free G-module with trivial action, then $H^1(G, A) = (0)$. For $H^1(G, A) = \operatorname{dir\,lim}_U H^1(G/U, A^U)$. As G acts trivially, example (7) shows that

$$H^1(G, A) = \operatorname{dir\,lim}_U \operatorname{Hom}(G/U, A).$$

But G/U is finite, and A is torsion free, so $\operatorname{Hom}(G/U, A) = (0)$. A particularly important case of this remark is: $H^1(G, Z) = (0)$.

If A is a uniquely divisible G-module (i.e., for every integer $m > 0$, the sequence $0 \to A \xrightarrow{m} A \to 0$ is exact), then $H^r(G, A)$ vanishes for $r > 0$. To see this one need only note that multiplication by m is an isomorphism for all m on the groups $H^r(G, A)$. If $r > 0$, these groups are torsion groups, so they must vanish. If A is a p-torsion module, the map $A \to A$ *via* multiplication by q, for a prime number $q \neq p$, is an isomorphism. It follows that $H^r(G, A)$ is a p-torsion group. Consequently, if G is a p-group, and A is a q-torsion module, the cohomology groups $H^r(G, A)$ vanish for positive r.

Let us apply these observations to the exact sequence $0 \to \mathbf{Z} \to \mathbf{Q} \to \mathbf{Q}/\mathbf{Z} \to 0$ of G-modules with trivial action. As $\mathbf{Q}$ is uniquely divisible, the cohomology sequence yields the *isomorphisms*

$$H^r(G, \mathbf{Q}/\mathbf{Z}) \xrightarrow{\ \delta\ } H^{r+1}(G, \mathbf{Z})$$

for all $r > 0$. When $r = 1$, $H^r(G, \mathbf{Q}/\mathbf{Z}) = \mathrm{Hom}_C(G, \mathbf{Q}/\mathbf{Z})$ = Pontrjagin dual of G^{ab}, where G^{ab} is $G/\overline{[G, G]}$ – the maximal abelian quotient of G. Consequently, the group $H^2(G, \mathbf{Z})$ is the Pontrjagin dual of G^{ab}. When G is abelian this shows that one can recover G from a knowledge of its cohomology with coefficients in $\mathbf{Z}$.

To study cup products, suppose we have two abelian categories $\mathfrak{A}, \mathfrak{B}$ respectively and two δ-functors $\{T^n\}$ from $\mathfrak{A}$ to abelian groups, $\{S^n\}$ from $\mathfrak{B}$ to abelian groups. We can make a new δ-functor from the abelian category $\mathfrak{A} \times \mathfrak{B}$ to abelian groups if we set

$$(8) \qquad (T \times S)^n(A \times B) = \coprod_{i+j=n} T^i(A) \times S^j(B) \ ,$$

and if we set $\delta_{T \times S}$ in dimension n equal to $\Sigma_{i+j=n} \Delta_{ij}$, where

$$\Delta_{ij}: T^i(A) \times S^j(B) \longrightarrow T^{i+1}(A) \times S^j(B) \oplus T^i(A) \times S^{j+1}(B)$$

via the formula

$$(9) \qquad \Delta_{ij} = \delta_T \times 1 + (-1)^i 1 \times \delta_S \ .$$

When $\mathfrak{A} = \mathfrak{B}$ and $T = S$, the Eilenberg-Zilber Theorem connects $T \times T$ and the δ-functor $A \times B \rightsquigarrow \{T^n(A \otimes B)\}_{n=0}^{\infty}$. This can be used to construct cup-products. For example, if A, B, C are G-modules for a profinite group G; let $\theta: A \times B \to C$ be a G-pairing. That is, θ is a bi-additive mapping and $\theta(\sigma a, \sigma b) = \sigma\theta(a, b)$. Then we get

THEOREM 9. *The* G-*pairing* θ *gives rise to a "cup-product"*

$$(10) \qquad \cup_\theta: H^r(G, A) \times H^s(G, B) \longrightarrow H^{r+s}(G, C)$$

for every $r, s \geqq 0$. *This cup-product is a* $\mathbf{Z}$*-bilinear mapping uniquely determined by its value in dimension zero and the following fact: Given three exact sequences*

$$(11)\quad \begin{cases} 0 \longrightarrow A' \longrightarrow A \longrightarrow A'' \longrightarrow 0 \\ 0 \longrightarrow B' \longrightarrow B \longrightarrow B'' \longrightarrow 0 \\ 0 \longrightarrow C' \longrightarrow C \longrightarrow C'' \longrightarrow 0 \end{cases}$$

such that $A \times B \to C$, $A'' \times B'' \to C''$, $A' \times B'' \to C'$, $A'' \times B' \to C'$ *under* θ, *for* $a \in H^r(A'')$, $b \in H^s(B'')$ *we have*

$$(12)\qquad \delta(a \cup b) = \delta a \cup b + (-1)^r a \cup \delta b \ .$$

Moreover, when $A = B$ *the cup-product is graded commutative, i.e.,*

$$(13)\qquad a \cup b = (-1)^{rs} b \cup a \ ,$$

and, in general, it is associative in the usual sense.

A complete proof of this result may be found in [CE, Ch. 12], however, remark 1 below provides an alternative description from which the reader may make his own proof.

Remarks. (1) One can make the cup-product explicit as follows: If $a \in H^r(G, A)$, $b \in H^s(G, B)$, let $f(\sigma_1, \ldots, \sigma_r)$ (resp. $g(\sigma_1, \ldots, \sigma_s)$) be a cocycle representing a (resp. b). Form the cochain

$$(f \cup g)(\tau_1, \ldots, \tau_{r+s}) = \theta(f(\tau_1, \ldots, \tau_r), \tau_1 \tau_2 \cdots \tau_r g(\tau_{r+1}, \ldots, \tau_{r+s})) \ ,$$

and observe that it is a cocycle. Then $a \cup b$ is the class of $f \cup g$.

(2) A G-pairing θ yields an S-pairing for any closed subgroup S of G. Of course, there is a corresponding cup-product for the cohomology over S, and the relationship between these cup-products is given by

PROPOSITION 11. *Let* S *be an open subgroup of the profinite group* G *and let* $A \times B \to C$ *be a* G*-pairing of the* G*-modules* A, B, C. *If* $a \in H^r(G, A)$, $b \in H^s(S, \pi^* B)$, *then*

(14) $$a \cup \mathrm{tr}(b) = \mathrm{tr}(\mathrm{res}\, a \cup b) \ .$$

That is, with respect to the pairing on cohomology, the mappings tr *and* res *are adjoint.*

Proof: Formula (14) is valid in dimension zero as one checks. The standard uniqueness argument now completes our proof. Q.E.D.

§4. A Utilitarian View of Spectral Sequences

We shall not give any comprehensive theory here—rather a small sampling of the basic theorems and definitions with emphasis on their use in the theory to be developed later on. For general complete treatments we may cite [CE, Go, EM].

Suppose A is an abelian group. A *filtration* on A is a family of subgroups $\{A^n\}_{n=-\infty}^{\infty}$ of A with $A^{n+1} \subseteq A^n$. We shall always assume:

$$A^{-\infty} = A, \qquad A^{\infty} = (0) \ .$$

(Here, as is usual, $A^{-\infty}$ signifies $\bigcup A^n$, and A^{∞} means $\bigcap A^n$.) The group A together with its filtration is a *filtered abelian group.* We can form its *associated graded group,* $\mathrm{gr}(A)$, as follows

$$\mathrm{gr}(A) = \coprod_{n=-\infty}^{\infty} \mathrm{gr}(A)_n; \qquad \mathrm{gr}(A)_n = A^n / A^{n+1} \ .$$

The usual case is when A is a complex such as $\{C^n(G,B)\}_{n=0}^{\infty}$ (G is profinite, B a G-module), then we set $A = \coprod_{n=0}^{\infty} C^n(G,B)$. We can filter A *via* $A^p = \coprod_{n \geq p} C^n(G,B)$. This filtration is *compatible with* the natural grading on $A\,(= C^*(G,B))$, namely, we have

(15) $$A^p = \coprod_q A^p \cap A_{p+q}$$

(Observe: $A_n = C^n(G,B)$ in this case.) Equation (15) insures that, when A is graded, the associated graded object induced by the filtration will be A again with the original grading.

Now our specific example $A = C^*(G, B)$ also has a natural "differentiation" $d = \amalg_n d_n$. Here, $d_n : C^n(G, B) \to C^{n+1}(G, B)$ is the coboundary map, so that d is an endomorphism of A of degree 1 (i.e., raises degrees by 1) and $d^2 = d \circ d$ vanishes. Moreover, the filtration $\{A^p\}$ is *compatible* with the differentiation, i.e., $d(A^p) \subseteq A^p$. It follows that A^p is a complex with the induced differentiation $d \mid A^p$.

Since we assume the gradation on our group A is compatible with its filtration, Equation (15) shows that A^p is also graded, being the direct sum of the groups $A^{p,q}$ where

(16) $$A^{p,q} = A^p \cap A_{p+q} \, .$$

Now observe with all these indices that: (a) We recover A^p if we sum the $A^{p,q}$ over all q, (b) the subgroup $A^{p,q}$ is a subgroup of the pth part of the filtration, A^p, (c) the subgroup $A^{p,q}$ is a subgroup of the $(p+q)^{th}$ homogeneous part, A_{p+q}, of A in its gradation. For these reasons, p is called the *filtration index*, $p+q$ is called the *total (or grading) index*, and q is the *complementary index*.

Let us look at gr(A). It is graded and also filtered in the obvious way and these are compatible. Looking more closely, we see that

$$\begin{aligned} A^p/A^{p+1} &= \amalg_q A^p \cap A_{p+q} \Big/ \amalg_q A^{p+1} \cap A_{p+q} \\ &= \amalg_q \left(A^p \cap A_{p+q} / A^{p+1} \cap A_{p+q}\right). \end{aligned}$$

If we set

(17) $$\mathrm{gr}(A)^{p,q} = A^{p,q}/A^{p+1,\,q-1} \, ,$$

then $\mathrm{gr}(A) = \amalg_{p,q} \mathrm{gr}(A)^{p,q}$ —so that gr(A) is bigraded. Again, we see the reasons behind the terms *filtration, complementary* and *total index.*

In the case of a graded group A, there is one class of compatible filtrations which is so important that we shall exclude *all* others from the discussion. These are the *regular filtrations.* A filtration is *regular* if and only if for every n, we can find $\mu(n)$, such that for all $p > \mu(n)$, we have $A^p \cap A_n = (0)$. (Observe that the canonical filtration on a graded A is always regular.)

We now suppose A is a graded, filtered group whose filtration is regular and compatible with the gradation. We also assume A possesses a differentiation d compatible with the filtration. Then d produces a *homology module* $H^*(A)$, graded by

$$H^*(A) = \coprod_n H^n(A); \qquad H^n(A) = \ker d_n / \operatorname{Im} d_{n-1} \ .$$

The inclusion mappings $A^p \to A$ induce maps of the respective homology modules $H^*(A^p) \to H^*(A)$ whose images, $H(A)^p$, form a filtration of $H^*(A)$. One checks that the filtration and gradation on $H^*(A)$ are compatible. This instantly gives us a bigradation on $H^*(A)$, namely

$$H^*(A) = \coprod_{p,q} H^*(A)^{p,q}, \text{ where } H^*(A)^{p,q} = H(A)^p \cap H^{p+q}(A) \ .$$

Collecting all the data given by such an A, we have

(α) $gr(A)$, bigraded by $gr(A)^{p,q} = A^{p,q}/A^{p+1,q-1}$
where $A^{p,q} = A^p \cap A_{p+q}$

(β) $H^*(A)$, bigraded by $H^*(A)^{p,q} = H(A)^p \cap H^{p+q}(A)$

(γ) $gr(H^*(A))$, bigraded by $gr(H^*(A))^{p,q} = H^*(A)^{p,q}/H^*(A)^{p+1,q-1}$
and

(δ) $H^*(gr(A))$, bigraded by $H^*(gr(A))^{p,q} = H^{p+q}(gr(A)_p)$
$= H^{p+q}(A^p/A^{p+1})$.

The central problem all this orgy of indices is meant to handle is: *Given* A *as above, compute* $H^*(A)$. Frequently, with relative ease, one can compute $H^*(gr(A))$, i.e., (δ). *The method of spectral sequences is an iterative procedure which passes from* $H^*(gr(A))$ *to* $gr(H^*(A))$. This is weaker than finding $H^*(A)$, but often is "just as good." Hence, the whole machinery of spectral sequences is an analysis of the "non-commutativity" of gr and H^*.

The iteration alluded to in the last paragraph is merely the repeated passage to homology ending in the "limit" with the group $gr(H^*(A))$. Formally, we have

DEFINITION 4. A (*cohomological*) *spectral sequence* is a system $E = \langle E_r^{p,q}, d_r^{p,q}, \alpha_r^{p,q}, E_n, \beta^{p,q} \rangle$ formed of

(a) Groups $E_r^{p,q}$ for $p,q \geqq 0$, $r \geqq 2$

(b) Homomorphisms $d_r^{p,q}: E_r^{p,q} \to E_r^{p+r,\, q-r+1}$

such that $d_r^{p+r,\, q-r+1} \circ d_r^{p,q} = 0$ for all p,q,r

(c) Isomorphisms $\alpha_r^{p,q}:(\ker d_r^{p,q} / \operatorname{Im} d_r^{p-r,\, q+r-1}) \to E_{r+1}^{p,q}$

(d) Groups E_n for $n \geqq 0$ filtered by a decreasing filtration, and

(e) Isomorphisms $\beta^{p,q}: E_\infty^{p,q} \to gr(E)^{p,q}$ where

$$gr(E)^{p,q} = E^p \cap E_{p+q} / E^{p+1} \cap E_{p+q}, \quad E = \coprod_n E_n,$$

$E^p = p^{th}$ filtration group of E, and $E_\infty^{p,q}$ is the common value of $E_r^{p,q}$ for very large r.

Observe in conjunction with (e) that from (a) and (c), we deduce that for $r > \sup(p, q+1)$, we have $E_r^{p,q} = E_{r+1}^{p,q} = \cdots$. Hence, $E_\infty^{p,q}$ is reached for $r > \sup(p, q+1)$. (Therefore, $E_2^{0,0} = E_\infty^{0,0}$, $E_2^{1,0} = E_\infty^{1,0}$.) The object $E = \coprod_n E_n$ is called the *ending of the spectral sequence*, and the whole definition is written in the compact form

$$E_2^{p,q} \underset{p}{\Longrightarrow} E$$

which means there exist $E_r^{p,q}$, d's, α's, β's, E, etc. so that Definition 4 is valid. Also the p under the arrow (usually omitted) denotes the filtration index.

THEOREM 10. *Let* A *be a complex having a regular filtration compatible with both the grading and differentiation. Then there exists a spectral* *sequence* $E_2^{p,q} \underset{p}{\Longrightarrow} H^*(A)$, *where* $H^*(A)$ *is the homology of* A *graded and filtered as described above and* $E_2^{p,q}$ *is the homology of* $H^*(\mathrm{gr}(A))$– *so that* $E_1^{p,q} = H^{p,q}(\mathrm{gr}(A)) = H^{p+q}(A^p/A^{p+1})$.

Remark. In the course of proving Theorem 10, we shall make heavy use of the following simple statement whose proof will be left as an exercise:

(L) *Let*

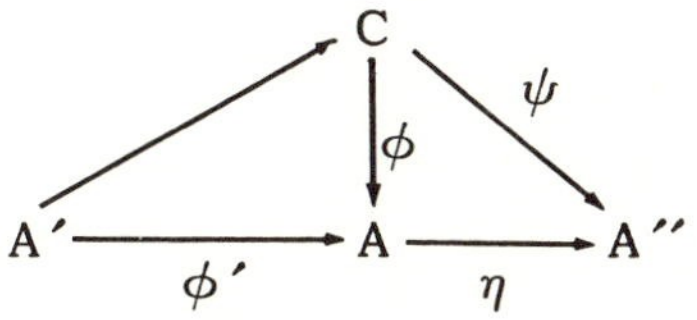

be a commutative diagram with exact bottom row. Then η *induces an isomorphism* $\mathrm{im}\,\phi/\,\mathrm{im}\,\phi' \xrightarrow{\sim} \mathrm{im}\,\psi$.

Proof of Theorem 10.† Consider the exact sequence

$$0 \longrightarrow A^p \longrightarrow A^{p-r+1} \longrightarrow A^{p-r+1}/\,A^p \longrightarrow 0 \ .$$

Upon applying cohomology, we obtain

$$\cdots \longrightarrow H^{p+q-1}(A^{p-r+1}) \longrightarrow H^{p+q-1}(A^{p-r+1}/A^p) \xrightarrow{\delta^*} H^{p+q}(A^p) \longrightarrow \cdots$$

There is also the natural map $H^{p+q}(A^p) \to H^{p+q}(A^p/A^{p+1})$ induced by the projection $A^p \to A^p/A^{p+1}$. Moreover, we have the projection $A^p/A^{p+r} \to A^p/A^{p+1}$, which induces a map on cohomology

$$H^{p+q}(A^p/A^{p+r}) \longrightarrow H^{p+q}(A^p/A^{p+1}) \ .$$

Set

$$(18) \qquad \begin{aligned} Z_r^{p,q} &= \mathrm{im}\,(H^{p+q}(A^p/A^{p+r}) \longrightarrow H^{p+q}(A^p/A^{p+1})) \\ B_r^{p,q} &= \mathrm{im}\,(H^{p+q-1}(A^{p-r+1}/A^p) \longrightarrow H^{p+q}(A^p/A^{p+1})), \end{aligned}$$

the latter map being the composition of δ^* and the projection.

† The reader is advised to skip this proof on a first reading.

The inclusion $A^{p-r+1} \subseteq A^{p-r}$ yields a map $A^{p-r+1}/A^p \to A^{p-r}/A^p$; hence, we obtain the inclusion relation $B_r^{p,q} \subseteq B_{r+1}^{p,q}$. In a similar way, the projection $A^p/A^{p+r+1} \to A^p/A^{p+r}$ yields the inclusion $Z_{r+1}^{p,q} \subseteq Z_r^{p,q}$. When $r = \infty$, the coboundary map yields the inclusion $B_\infty^{p,q} \subseteq Z_\infty^{p,q}$; consequently, we can write:

$$\cdots \subseteq B_r^{p,q} \subseteq B_{r+1}^{p,q} \subseteq \cdots \subseteq B_\infty^{p,q} \subseteq Z_\infty^{p,q} \subseteq \cdots \subseteq Z_{r+1}^{p,q} \subseteq Z_r^{p,q} \subseteq \cdots .$$

Set $E_r^{p,q} = Z_r^{p,q}/B_r^{p,q}$, $1 \leqq r \leqq \infty$. Then, when $r = 1$, $B_1^{p,q} = (0)$ and

$$Z_1^{p,q} = H^{p+q}(A^p/A^{p+1}) \ ;$$

so we obtain $E_1^{p,q} = H^{p+q}(A^p/A^{p+1}) = H^{p,q}(\mathrm{gr}(A))$. On the other hand, when $r = \infty$,

$$Z_\infty^{p,q} = \mathrm{im}\,(H^{p+q}(A^p) \longrightarrow H^{p+q}(A^p/A^{p+1}))$$

$$B_\infty^{p,q} = \mathrm{im}\,(H^{p+q-1}(A/A^p) \longrightarrow H^{p+q}(A^p/A^{p+1})) \quad .$$

Now the exact sequence $0 \to A^p/A^{p+1} \to A/A^{p+1} \to A/A^p \to 0$ yields the cohomology sequence

$$\cdots \longrightarrow H^{p+q-1}(A/A^p) \xrightarrow{\delta^*} H^{p+q}(A^p/A^{p+1}) \longrightarrow H^{p+q}(A/A^{p+1}) \longrightarrow \cdots$$

and the exact sequence $0 \to A^p \to A \to A/A^p \to 0$ gives rise to the connecting homomorphism $H^{p+q-1}(A/A^p) \to H^{p+q}(A^p)$. Consequently, we obtain the commutative diagram (with exact bottom row)

$$\begin{array}{ccccc} & & H^{p+q}(A^p) & & \\ & \nearrow & \downarrow & \searrow & \\ H^{p+q-1}(A/A^p) & \longrightarrow & H^{p+q}(A^p/A^{p+1}) & \longrightarrow & H^{p+q}(A/A^{p+1}) \end{array} ,$$

and (L) yields an isomorphism

$$\xi^{p,q}\colon E_\infty^{p,q} = Z_\infty^{p,q}/B_\infty^{p,q} \longrightarrow \mathrm{im}\,(H^{p+q}(A^p) \longrightarrow H^{p+q}(A/A^{p+1})) \ .$$

But another application of (L) to the diagram

$$\begin{array}{ccccc} & \nearrow & H^{p+q}(A^p) & \searrow & \\ & & \downarrow & & \\ H^{p+q}(A^{p+1}) & \longrightarrow & H^{p+q}(A) & \longrightarrow & H^{p+q}(A/A^{p+1}) \end{array}$$

gives us the isomorphism

$$\eta^{p,q}:\ gr^{p,q}(H(A)) \longrightarrow \operatorname{im}(H^{p+q}(A^p) \longrightarrow H^{p+q}(A/A^{p+1}))\ .$$

Thus, $(\eta^{p,q})^{-1} \circ \xi^{p,q}$ is the isomorphism $\beta^{p,q}$ required by part (e) of Definition 4.

Only two things remain to be proven to complete the proof of Theorem 10. They are the verification of (b) and (c) of Definition 4, and the observation that E_∞ as defined in Definition 4 is the same E_∞ computed above. The verification of (b) and (c) depends upon (L). Specifically, we have the two commutative diagrams (with obvious origins)

$$\begin{array}{ccccc} & \nearrow & H^{p+q}(A^p/A^{p+r}) & \overset{\theta}{\searrow} & \\ & & \downarrow & & \\ H^{p+q}(A^p/A^{p+r+1}) & \longrightarrow & H^{p+q}(A^p/A^{p+1}) & \underset{\delta^*}{\longrightarrow} & H^{p+q+1}(A^{p+1}/A^{p+r+1}) \end{array}$$

$$\begin{array}{ccccc} & \nearrow & H^{p+q}(A^p/A^{p+r}) & \overset{\theta}{\searrow} & \\ & & \downarrow & & \\ H^{p+q}(A^{p+1}/A^{p+r}) & \underset{\delta^*}{\longrightarrow} & H^{p+q+1}(A^{p+r}/A^{p+r+1}) & \longrightarrow & H^{p+q+1}(A^{p+1}/A^{p+r+1}) \end{array}\ .$$

Here, the map θ is the composition

$$H^{p+q}(A^p/A^{p+r}) \longrightarrow H^{p+q+1}(A^{p+r}) \longrightarrow H^{p+q+1}(A^{p+1}/A^{p+r+1})\ .$$

Now (L) yields the following facts:

$$Z_r^{p,q}/Z_{r+1}^{p,q} \xrightarrow{\ \approx\ } \operatorname{im}\theta\ ,$$

$$B_{r+1}^{p+r,\,q-r+1}/B_r^{p+r,\,q-r+1} \xrightarrow{\ \sim\ } \operatorname{im}\theta \ ,$$

that is,

$$\delta_r^{p,q}\colon Z_r^{p,q}/Z_{r+1}^{p,q} \xrightarrow{\ \sim\ } B_{r+1}^{p+r,\,q-r+1}/B_r^{p+r,\,q-r+1} \ .$$

As $B_r^{p,q} \subseteq Z_s^{p,q}$ for every r and s, there is a surjection

$$\pi_r^{p,q}\colon E_r^{p,q} \longrightarrow Z_r^{p,q}/Z_{r+1}^{p,q}$$

with kernel $Z_{r+1}^{p,q}/B_r^{p,q}$; and there exists an injection

$$\sigma_{r+1}^{p+r,\,q-r+1}\colon B_{r+1}^{p+r,\,q-r+1}/B_r^{p+r,\,q-r+1} \longrightarrow E_r^{p+r,\,q-r+1} \ .$$

The composition $\sigma_{r+1}^{p+r,\,q-r+1} \circ \delta_r^{p,q} \circ \pi_r^{p,q}$ is the map $d_r^{p,q}$ from $E_r^{p,q}$ to $E_r^{p+r,\,q-r+1}$ required by (b). Observe that,

$$\operatorname{im} d_r^{p-r,\,q+r-1} = B_{r+1}^{p,q}/B_r^{p,q} \subseteq Z_{r+1}^{p,q}/B_r^{p,q} = \ker d_r^{p,q} \ ;$$

hence,

$$H^{p,q}(E_r) = \ker d_r^{p,q}/\operatorname{im} d_r^{p-r,\,q+r-1} = Z_{r+1}^{p,q}/B_{r+1}^{p,q} = E_{r+1}^{p,q} \ ,$$

as required by (c).

To prove that $E_\infty^{p,q}$ as defined above is the common value of $E_r^{p,q}$ for large enough r, we must make use of the regularity of our filtration. Consider then the commutative diagram

$$\begin{array}{ccccc} & \nearrow & H^{p+q}(A^p/A^{p+r}) & \overset{\lambda}{\searrow} & \\ & & \downarrow & & \\ H^{p+q}(A^p) & \longrightarrow & H^{p+q}(A^p/A^{p+1}) & \longrightarrow & H^{p+q+1}(A^{p+1}) \ , \end{array}$$

where λ is the composition

$$H^{p+q}(A^p/A^{p+r}) \xrightarrow{\ \delta^*\ } H^{p+q+1}(A^{p+r}) \longrightarrow H^{p+q+1}(A^{p+1})$$

By (L), $Z_r^{p,q}/Z_\infty^{p,q} \xrightarrow{\ \sim\ } \operatorname{im}\lambda$. However, if $r > \mu(p+q+1)-p$, then δ^* is the zero map. This shows $\operatorname{im}\lambda = (0)$; hence, we have proven

$$Z_r^{p,q} = Z_\infty^{p,q} \quad \text{for} \quad r > \mu(p+q+1)-p \ .$$

It is easy to see that $\bigcup_r B_r^{p,q} = B_\infty^{p,q}$; hence, we obtain maps

$$E_r^{p,q} = Z_r^{p,q}/B_r^{p,q} \longrightarrow Z_s^{p,q}/B_s^{p,q} = E_s^{p,q}$$

for $s \geqq r > \mu(p+q+1)-p$, and these maps are surjective. (The maps are in fact induced by the $d_r^{p-r,\, q+r-1}$'s because of the equality

$$E_r^{p,q}/\mathrm{im}\ d_r^{p-r,\, q+r-1} = (Z_r^{p,q}/B_r^{p,q})/(B_{r+1}^{p,q}/B_r^{p,q}) = E_{r+1}^{p,q}$$

for $r > \mu(p+q+1)-p$.) Obviously, the direct limit of the mapping family

$$E_r^{p,q} \longrightarrow E_{r+1}^{p,q} \longrightarrow \cdots \longrightarrow E_s^{p,q} \longrightarrow \cdots$$

is the group $Z_\infty^{p,q}/\bigcup B_r^{p,q} = E_\infty^{p,q}$, and this completes the proof. Q.E.D.

From our present point of view, the uses of Theorem 10 are more important than its proof. In this line, we shall make certain remarks intended to be of service in applying Theorem 10 and in understanding the spectral sequence in general.

The definition of $E_\infty^{p,q}$ shows that

$$E_\infty^{p,q} = H^{p+q}(A) \cap H^*(A)^p / H^{p+q}(A) \cap H^*(A)^{p+1} \ .$$

Therefore, for $p+q = n$, the $E_\infty^{p,q}$ are the composition factors in the filtration

$$H^n(A) \supseteq H^n(A)^1 \supseteq H^n(A)^2 \supseteq \cdots \supseteq H^n(A)^v \supseteq \cdots$$

The group $E_\infty^{p,n-p}$ is the pth composition factor in $H^n(A)$. A pictorial representation of the above situation is very convenient. In this representation, the groups $E_r^{p,q}$ (for fixed r) are represented as points in the pq plane, *viz:*

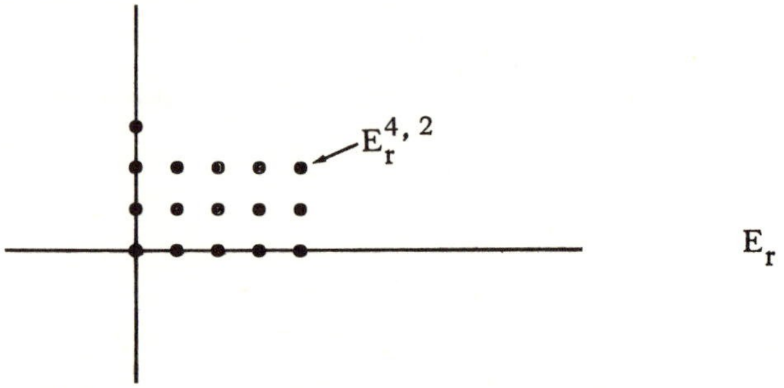

and the differentiation $d_r^{p,q}$ is represented as an arrow "going over r and down r − 1." So the situation discussed above may be represented

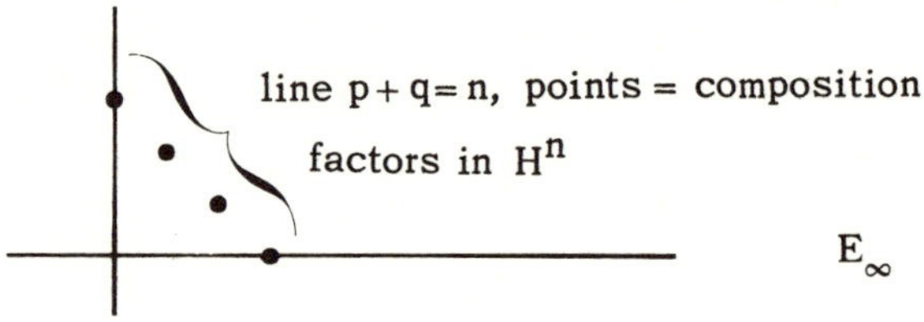

Our arguments show that, since $E_\infty^{p,q} = (0)$ for $q < 0$, one has

$$H^n(A)^{n+1} = H^n(A)^{n+2} = \cdots \quad .$$

However, the filtration is regular, so we deduce $H^n(A)^v = (0)$ for sufficiently large v. Therefore, $H^n(A)^v = (0)$ for $v \geqq n+1$, and we have

$$(19) \quad \begin{cases} E_\infty^{n,0} = H^n(A)^n \subseteq H^n(A); \quad E_\infty^{n-1,1} = H^n(A)^{n-1}/H^n(A)^n, \ldots \\ E_\infty^{0,n} = H^n(A)/H^n(A)^1, \text{ a homomorphic image of } H^n(A). \end{cases}$$

As an example, suppose we could show $E_\infty^{p,q} = (0)$ for $p+q > n$ and all p. It would follow that $H^r(A) = (0)$ for every $r > n$.

We shall say a spectral sequence *degenerates*, or *is degenerate*, if and only if for every n, there is an integer q(n) such that

$$E_2^{n-q,q} = (0) \quad \text{for all } q \neq q(n) \ .$$

In this case, clearly $E_\infty^{n-q,q} = (0)$ for all $q \neq q(n)$; therefore, there is only one composition factor, the $n-q^{\text{th}}$, for $q = q(n)$. This yields

$$H^n(A) \xrightarrow{\sim} E_\infty^{n-q(n),\,q(n)}$$

when the spectral sequence degenerates. The most common case is when $q(n) = 0$ for all n, i.e., $E_2^{p,q} = (0)$ for all $q > 0$. Then the isomorphisms become

$$E_2^{p,0} \cong H^p(A), \quad \text{for every } p.$$

Now consider a spectral sequence $E_2^{p,q} \underset{p}{\Longrightarrow} H^*(A)$. Since $E_2^{1,0} = E_\infty^{1,0} = H^1(A)^1 \subseteq H^1(A)$, we have the monomorphism (called an *edge homomorphism*)

$$0 \longrightarrow E_2^{1,0} \longrightarrow H^1(A) .$$

Now $E_2^{0,1}$ has homology $E_3^{0,1}$, etc., but $E_3^{0,1} = E_\infty^{0,1}$, as we know. So let us examine the homology $H(E_2^{0,1})$ more closely.

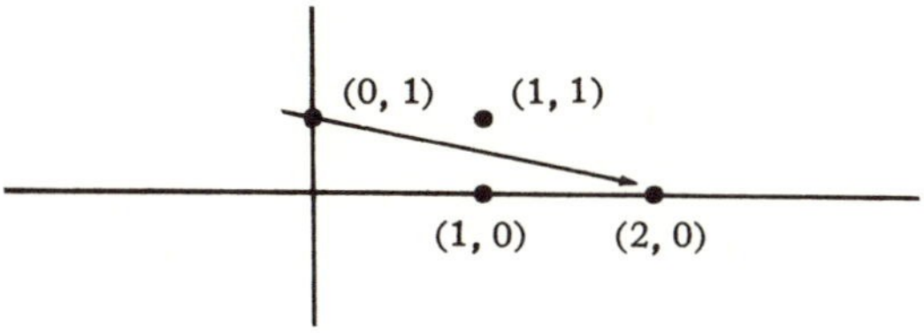

$d_2^{0,1}\colon E_2^{0,1} \to E_2^{2,0}$, as shown and its kernel is $Z_2^{0,1}$. But $B_2^{0,1}$ is zero, so $E_\infty^{0,1} = E_3^{0,1} = \ker d_2^{0,1} = Z_2^{0,1}$. Now $E_\infty^{0,1}$ is $H^1(A)/H^1(A)^1$, so there is a surjection $H^1(A) \to E_\infty^{0,1}$, with kernel $H^1(A)^1 = E_\infty^{1,0} = E_2^{1,0}$. But according to the above, $E_\infty^{0,1} \to E_2^{0,1} \to E_2^{2,0}$ is exact; consequently, we get

(*) $$0 \longrightarrow E_2^{1,0} \longrightarrow H^1(A) \longrightarrow E_2^{0,1} \xrightarrow{d_2^{0,1}} E_2^{2,0}$$

is exact. What of $E_2^{2,0}$? Since $3 > \sup(2, 0)$, $E_3^{2,0} = E_\infty^{2,0} = H^2(A)^2$ which is embedded in $H^2(A)$. Now $d_2^{2,0}$ is the zero map, so $Z_3^{2,0} = \ker d_2^{2,0} = E_2^{2,0}$, and $B_3^{2,0} = \operatorname{Im} d_2^{0,1}$. Thus the projection, $E_2^{2,0} = Z_3^{2,0} \to E_3^{2,0} = E_\infty^{2,0}$ has kernel $\operatorname{Im} d_2^{0,1}$, and we finally see that, coupled with the injection $E_\infty^{2,0} \hookrightarrow H^2(A)$, we get the exact sequence

(**) $$E_2^{0,1} \xrightarrow{d_2^{0,1}} E_2^{2,0} \longrightarrow H^2(A) .$$

And now, if we put (*) and (**) together, we obtain

PROPOSITION 12. *If* $E_2^{p,q} \underset{p}{\Longrightarrow} H^*(A)$, *then there is the exact sequence of terms of low degree*

$$(20)\quad 0 \longrightarrow E_2^{1,0} \longrightarrow H^1(A) \longrightarrow E_2^{0,1} \xrightarrow{d_2^{0,1}} E_2^{2,0} \longrightarrow H^2(A)\ .$$

Even more is true. *Namely if* $E_2^{p,q} = (0)$ *for* $0 < q < n$, *then one has the exact sequence of terms of low degree*

$$(21)\quad 0 \longrightarrow E_2^{n,0} \longrightarrow H^n(A) \longrightarrow E_2^{0,n} \longrightarrow E_2^{n+1,0} \longrightarrow H^{n+1}(A)$$

and

$$(22)\qquad E_2^{p,0} \cong H^p(A) \quad \text{if} \quad 0 \leqq p \leqq n-1\ .$$

We omit the proof of this. In the case of interest to us, it follows from Proposition 12 by a simple dimension shifting argument. The general case is fairly technical.

The most important method for generating spectral sequences is the following: Suppose that F, G are functors whose composition $F \circ G = \Gamma$ exists. Assume that both F and G are left-exact so that Γ is also left-exact, and assume that all categories considered possess enough injectives —so that $R^n\Gamma$, R^pF, R^qG all exist. Lastly, assume the *crucial hypothesis*: G *takes injectives into* F*-acyclic objects*, i.e., $R^pF(GQ) = (0)$ for $p > 0$ and Q injective. Then there exists a spectral sequence for every object A of the category where G is defined

$$(23)\qquad R^pF(R^qG(A)) \underset{p}{\Longrightarrow} R^*\Gamma(A)$$

called the *spectral sequence for composed functors*.

There are two noteworthy applications of Equation (23).

(1) Let $\underline{G}$ be a profinite group, S a closed subgroup, and A an S-module. Consider the functors $\Gamma: A \leadsto A^S$, $G: A \leadsto \pi_*A$ and $F: B \leadsto B^{\underline{G}}$. We know that $\Gamma = F \circ G$; hence,

$$H^p(\underline{G}, R^q\pi_*A) \implies R^*\Gamma = H^*(S, A) \ .$$

This is called the *Leray Spectral Sequence of the map* $\pi: S \to \underline{G}$.

Now π_* is exact; hence, $R^q\pi_* = (0)$ for $q > 0$. The Leray Spectral Sequence thus degenerates, and we obtain the isomorphisms

$$E_2^{p,0} = H^p(\underline{G}, \pi_*A) \xrightarrow{\sim} H^p(S, A)$$

which is the Lemma of Shapiro. This is the idea we had in mind and to which we alluded in the proof of Proposition 5.

(2) Let $\underline{G}$ be a profinite group, let N be a closed, normal subgroup of $\underline{G}$, and let A be a $\underline{G}$-module. Let $\Gamma: A \rightsquigarrow A^{\underline{G}}$, $G: A \rightsquigarrow A^N$, and $F: B \rightsquigarrow B^{\underline{G}/N}$ (on $\underline{G}/N$-modules B). Then $R^qG(A) = H^q(N, A)$, $R^pF(B) = H^p(\underline{G}/N, B)$, and $\Gamma = F \circ G$. Consequently, we obtain the *Hochschild-Serre Spectral Sequence* [HS]

$$H^p(\underline{G}/N, H^q(N, A)) \underset{p}{\implies} H^*(\underline{G}, A) \tag{24}$$

Remarks. The existence of the spectral sequence (24) implies a continuous action of $\underline{G}/N$ on $H^q(N, A)$ for each q. This is given by choosing $\alpha \in H^q(N, A)$, $\bar{\tau} \in \underline{G}/N$ and representing them by a cocycle $f(\sigma_1, \dots, \sigma_q)$ and an element $\tau \in \underline{G}$, then by setting

$$f^\tau(\sigma_1, \dots, \sigma_q) = \tau f(\tau^{-1}\sigma_1\tau, \dots, \tau^{-1}\sigma_q\tau) \ .$$

One checks that f^τ is a cocycle whose class is independent of the choices of τ and f representing $\bar{\tau}$ and α respectively, and moreover that N acts trivially on the cohomology classes and that the induced $\underline{G}/N$ action is continuous. The exact sequence of terms of low degree yields

$$0 \to H^1(\underline{G}/N, A^N) \xrightarrow{\theta_1} H^1(\underline{G}, A) \xrightarrow{\theta_2} H^1(N, A)^{\underline{G}/N} \to H^2(\underline{G}/N, A^N) \xrightarrow{\theta_3} H^2(\underline{G}, A) \tag{25}$$

Again, checking shows that θ_1, θ_3 are inflation, and θ_2 is restriction. The map $d_2^{0,1}$ is called *transgression,* and (25) is usually called the *inflation-restriction* sequence. One can characterize tg (the transgression) as follows: $\beta = \mathrm{tg}(\alpha)$ if and only if there exists a 1-cochain f whose restriction to N is a cocycle representing α and whose coboundary δf is the inflation of a cocycle representing β. We shall see many applications of the Hochschild-Serre sequence later on.

CHAPTER III

COHOMOLOGICAL DIMENSION

§1. Definition and elementary properties.

In this chapter G will always denote a profinite group.

DEFINITION 5. The *cohomological dimension of* G *is less than or equal to* n, *written* $\mathrm{cd}\, G \leqq n$, if and only if $H^r(G, A) = (0)$ for all $r > n$ and all torsion G-modules A. The *strict cohomological dimension of* G *is less than or equal to* n ($\mathrm{scd}\, G \leqq n$) iff $H^r(G, A) = (0)$ for all $r > n$ and all G-modules A. If p is a prime number, the p-*(strict) cohomological dimension of* G *is* $\leqq n$, written $(\mathrm{s})\mathrm{cd}_p\, G \leqq n$, iff $H^r(G, A; p) = (0)$ for (all) resp. all torsion G-modules A and all $r > n$. The infimum over all n, such that $\mathrm{cd}\, G \leqq n$ (resp. $\mathrm{scd}\, G \leqq n$, $\mathrm{cd}_p\, G \leqq n$, $\mathrm{scd}_p\, G \leqq n$) is called the *cohomological dimension* (resp. *strict cohomological dimension,* p-*cohomological dimension, strict* p-*cohomological dimension*) *of* G. It is denoted $\mathrm{cd}\, G = n$, etc.

Obviously, one has

(a) $\mathrm{cd}\, G = \sup\limits_p \mathrm{cd}_p\, G$

(b) $\mathrm{scd}\, G = \sup\limits_p \mathrm{scd}_p\, G$

(c) $\mathrm{cd}\, G \leqq \mathrm{scd}\, G$, $\mathrm{cd}_p\, G \leqq \mathrm{scd}_p\, G$.

PROPOSITION 13. *For every prime number* p *we have*

$$\mathrm{cd}_p\, G \leqq \mathrm{scd}_p\, G \leqq \mathrm{cd}_p\, G + 1$$

$$\mathrm{cd}\, G \leqq \mathrm{scd}\, G \leqq \mathrm{cd}\, G + 1.$$

Consequently, cohomological and strict cohomological dimension are simultaneously finite or infinite.

Proof: Let A be a G-module and let tA be its torsion submodule. Then $0 \to tA \to A \to A/tA \to 0$ is exact, and A/tA is torsion-free. If we apply cohomology to the sequence, we get

$$\ldots \to H^r(G, tA) \to H^r(G, A) \to H^r(G, A/tA) \to \ldots$$

is exact. If we show that $H^r(G, A/tA; p) = (0)$ for $r > n+1$ (where $n = cd_p G$), then it will follow that $H^r(G, A) = (0)$ for $r > n+1$; hence, $scd_p G \leqq n+1$. Let $B = A/tA$. As B is torsion-free, we have the exact sequence

$$0 \to B \xrightarrow{p} B \to B/pB \to 0 \,.$$

This gives us

$$\ldots \to H^{r-1}(G, B/pB) \to H^r(G, B) \xrightarrow{p} H^r(G, B) \to \ldots$$

If $r > n+1$, then $r-1 > n$, and as B/pB is p-torsion, we deduce $H^{r-1}(G, B/pB) = (0)$ for $r > n+1$. But then multiplication by p is a monomorphism on $H^r(G,B)$ —and this shows that $H^r(G, B; p)$ vanishes for $r > n+1$, as contended. Thus $cd_p G \leqq scd_p G \leqq cd_p G + 1$. Upon passing to the sup over p, we get $cd\, G \leqq scd\, G \leqq cd\, G + 1$. Q.E.D.

PROPOSITION 14. *Let* A *be a* p*-divisible* G*-module, and assume* $cd_p G = n$. *Then* $H^n(G, A)$ *is* p*-divisible and* $H^{n+1}(G, A; p)$ *is trivial (even if* A *is not torsion). If* A *is not* p*-divisible but* $cd_p G = n$, *then the above statements hold with* n *replaced by* $n+1$.

Proof: We have the exact sequence $0 \to A_p \to A \xrightarrow{p} A \to 0$, where $A_p = \{a \in A \mid pa = 0\}$. The cohomology sequence gives us

$$H^n(G, A) \xrightarrow{p} H^n(G, A) \longrightarrow H^{n+1}(G, A_p) = (0)$$

is exact. Thus, $H^n(G, A)$ is p-divisible. Continuing the sequence, we obtain

$$(0) = H^{n+1}(G, A_p) \longrightarrow H^{n+1}(G, A) \xrightarrow{p} H^{n+1}(G, A) \,.$$

As before, this proves $H^{n+1}(G, A; p) = (0)$. In the general case, one has

the commutative diagram with exact rows

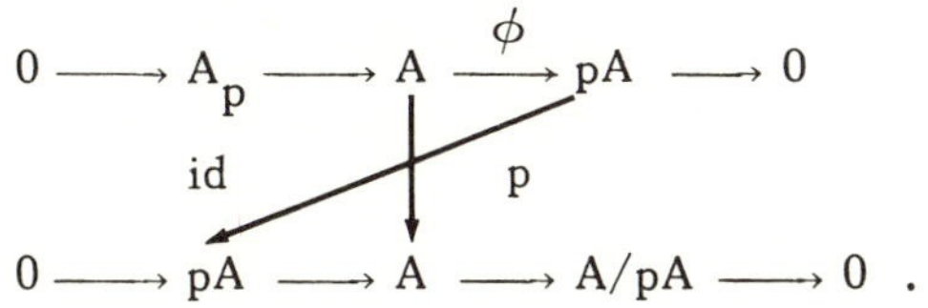

Proposition 13 shows that $\mathrm{scd}_p G \leqq n+1$, so we obtain the exact sequences

$$0 \longrightarrow H^{n+1}(G, A) \xrightarrow{\phi_*} H^{n+1}(G, pA) \longrightarrow 0$$

$$H^{n+1}(G, pA) \longrightarrow H^{n+1}(G, A) \longrightarrow 0 \ .$$

Together, they yield the p-divisibility of $H^{n+1}(G, A)$; the rest of the proposition is trivial. Q.E.D.

COROLLARY 1. *If* $\mathrm{cd}\, G = n$, *then for any* G-*module* A, $H^{n+1}(G, A)$ *is a divisible group.*

COROLLARY 2. *If* $\mathrm{cd}\, G = 1$, *then* G^{ab} *is torsion free.*

Proof: By Corollary 1, $H^2(G, Z)$ is divisible; so G^{ab} is the Pontrjagin dual of the divisible group $H^2(G, Z)$. Thus, G^{ab} is torsion free. Q.E.D.

THEOREM 11. $\mathrm{cd}_p G$ *is the smallest integer* $n \geqq 0$ *for which* $H^{n+1}(G, E) = (0)$ *for all finite, simple* G-*modules of* p-*power order.*

Proof: Let n be the integer in question, we know $n \leqq \mathrm{cd}_p G$ and all we need to show is $\mathrm{cd}_p G \leqq n$. Now given a G-module A, which is a torsion module, we may write $A = \mathrm{dir\ lim}\ A_\alpha$ where A_α ranges over the finitely generated G-submodules of A. However, a G-module is f.g. iff it is f.g. as Z-module. As A is a torsion module, the A_α are *finite* G-modules. The cohomology commutes with direct limits; hence, we are reduced to proving $H^{n+1}(G, A; p) = (0)$ for every finite G-module A. We can write $A = A_p \oplus A'$ where A' is the prime-to-p part of A, and A_p is the p-power part of A; then $H^r(G, A'; p)$ is trivial, so that $H^r(G, A_p) = H^r(G, A; p)$.

Consequently, we may assume $A = A_p$, i.e., A is a p-power module. For such a module there is a composition series

$$A > A_1 > A_2 > \cdots > A_m = (0)$$

whose factors are simple G-modules of p-power order. Now induction on the length of a composition series, and the cohomology sequence imply that if $H^{n+1}(G, E) = (0)$ for simple, p-power E, then $H^{n+1}(G, A) = (0)$ for all finite p-power A. This is what we needed, so the proof is complete. Q.E.D.

PROPOSITION 15. *Let* S *be a closed subgroup of* G, *and let* p *be any prime number. Then,* $\operatorname{cd}_p S \leqq \operatorname{cd}_p G$; $\operatorname{scd}_p S \leqq \operatorname{scd}_p G$. *We have equality if* $(G:S)$ *is prime-to-*p. *In particular,* $\operatorname{cd}_p G = \operatorname{cd}_p(G_p) = \operatorname{cd}(G_p)$, *where* G_p *is a* p*-Sylow subgroup of* G.

Proof: Let $\operatorname{cd}_p G = n$, and let B be a torsion S-module. If $\pi : S \to G$ is the inclusion, then Shapiro's Lemma yields the isomorphism $H^r(S, B; p) \xrightarrow{\sim} H^r(G, \pi_* B; p)$. The latter group vanishes if $r > n$, as $\pi_* B$ is again a torsion module. Consequently, $\operatorname{cd}_p S \leqq n$, as contended. The same argument works for scd_p as well.

Now suppose $(G:S)$ is prime-to-p. If A is any (torsion) G-module, then

$$\text{res}: H^r(G, A; p) \longrightarrow H^r(S, \pi^* A; p)$$

is injective by Proposition 10. If $r > (s)\operatorname{cd}_p S$, the right hand group is trivial, and this completes the proof. Q.E.D.

PROPOSITION 16. $\operatorname{cd} G = 0 \iff G = \{1\}$

$$\operatorname{cd}_p G = 0 \iff G_p = \{1\}.$$

Proof: ($\Longleftarrow$) This is trivial since the higher cohomology groups of the trivial group vanish.

($\Longrightarrow$) We need only prove this implication for the groups G_p. (For then $\operatorname{cd} G = 0$ implies $\operatorname{cd}_p G = 0$; hence, $G_p = \{1\}$ for all p, which yields $G = \{1\}$.) Thus we may and do assume G is a p-group (as $\operatorname{cd}_p G = \operatorname{cd}(G_p)$). Were $G_p \neq \{1\}$, there would exist an open normal subgroup $U \neq G_p$. Consequently,

G_p/U is a non-trivial finite p-group. As such, there exists a *non-trivial* homomorphism $G_p/U \to Z/pZ$ (exercise). This homomorphism is a non-trivial element of $H^1(G_p/U, Z/pZ)$ by example (7) and equation (7). However, we have the first part of the inflation-restriction sequence

$$0 \to H^1(G_p/U, Z/pZ) \xrightarrow{\text{inf}} H^1(G_p, Z/pZ) ,$$

and it shows that $H^1(G_p, Z/pZ) \neq (0)$. This contradicts our hypothesis, and the proof is complete. Q.E.D.

If G is a p-group, it is an amazing fact that its cohomological dimension is controlled by *one, single* module, and that this module is Z/pZ with *trivial* action. This result, which we shall now prove, is invaluable in establishing the exact value of the cd of certain groups. It is a very simple corollary of the following basic fact:

PROPOSITION 17. *If* G *is a profinite* p*-group, then the only finite, simple* G*-module of* p*-power order is* Z/pZ *with trivial action.*

Proof: Let E be a finite, simple G-module of p-power order. Given $x, y \in E$, we say that x is equivalent to y iff there is a $\sigma \in G$ with $\sigma x = y$. Clearly this notion is an equivalence relation on E, and the equivalence class of x is the orbit, $\mathcal{O}r(x)$, of x under G. Let U_x be the stabilizer of x in G, then one knows that the cardinal of $\mathcal{O}r(x)$ is exactly $(G : U_x)$. If $|\cdot|$ means the cardinal of $\cdot$, then

$$|E| = |E^G| + \sum_{\substack{\text{classes with} \\ \text{more than one} \\ \text{element}}} (G : U_x) \quad .$$

The summands on the far right side of the above equation are all p-powers as G is a p-group, and all are greater than 1. Consequently, as $p \mid |E|$, we deduce that $|E^G| > 1$. But E^G is a submodule of E, and E is simple. Therefore, $E^G = E$, and we have proved E has trivial action! And now to say E is simple means merely that E is a simple, abelian group of p-power order; that is, $E = Z/pZ$. Q.E.D.

THEOREM 12. *If* G *is a profinite* p-*group, then a necessary and sufficient condition that* $\mathrm{cd}_p G \leq n$ *is that* $H^{n+1}(G, \mathbf{Z}/p\mathbf{Z}) = (0)$. *Here,* $\mathbf{Z}/p\mathbf{Z}$ *is a* G-*module with trivial action.*

Proof: Combine Proposition 17 and Theorem 11. Q.E.D.

COROLLARY. *There is no profinite group for which* $\mathrm{scd}_p G = 1$ *for any prime number* p.

Proof: If $\mathrm{scd}_p G = 1$, then Proposition 13 shows that $\mathrm{cd}_p G = 0$ or 1. However, Proposition 16 shows that $\mathrm{cd}_p G = 0$ is impossible; hence, $\mathrm{cd}_p G = 1$. By Theorem 12, $H^1(G, \mathbf{Z}/p\mathbf{Z}) \neq (0)$. (We may assume $G = G_p$, because $\mathrm{scd}_p G = \mathrm{scd}_p G_p$.) Now consider the exact sequence

$$0 \longrightarrow \mathbf{Z} \xrightarrow{\ p\ } \mathbf{Z} \longrightarrow \mathbf{Z}/p\mathbf{Z} \longrightarrow 0$$

and apply cohomology. We obtain

$$\cdots \to H^1(G, \mathbf{Z}) \to H^1(G, \mathbf{Z}/p\mathbf{Z}) \to H^2(G, \mathbf{Z}) \to \cdots$$

is exact. But $H^1(G, \mathbf{Z}) = (0)$ (cf. remarks after Proposition 10); hence, $H^1(G, \mathbf{Z}/p\mathbf{Z})$ is contained in $H^2(G, \mathbf{Z})$. Thus, $H^2(G, \mathbf{Z}; p)$ is non-trivial, a contradiction since we assumed $\mathrm{scd}_p = 1$. Q.E.D.

There is a beautiful generalization of the Corollary above due to Serre. Here it is.

PROPOSITION 18. *Let* $\mathrm{cd}_p G = n$, *then a necessary and sufficient condition that* $\mathrm{scd}_p G = n$, *as well, is that*

$$H^{n+1}(S, \mathbf{Z}; p) = (0)$$

for every open subgroup S *of* G.

Proof: ($\Longrightarrow$) is clear, we need prove only ($\Longleftarrow$). Let B be any G-module; write $B = \text{dir lim}_\alpha B_\alpha$, with B_α a f.g. G-module. Because cohomology commutes with direct limits of coefficient modules we may assume B is itself f.g. If $b_1, \ldots, b_m$ are generators, each is left fixed by an open subgroup S_j of G, and all of B is left fixed by the open subgroup $S = \bigcap_j S_j$. Thus B as S-module (namely π^*B, where $\pi : S \to G$) is merely B as Z-module. Consequently, we can find an exact sequence

$$0 \longrightarrow K \longrightarrow Z^m \longrightarrow \pi^*B \longrightarrow 0$$

(m = rank of B as Z-module) in which K is a torsion module. Now apply cohomology over S, we get

$$\cdots \to H^{n+1}(S, K; p) \to H^{n+1}(S, Z; p)^m \longrightarrow H^{n+1}(S, \pi^*B; p) \to H^{n+2}(S, K; p) .$$

But $cd_p S \leqq n$, so the extremes of the sequence vanish. Also, by hypothesis, $H^{n+1}(S, Z; p) = (0)$ —and we deduce: $H^{n+1}(S, \pi^*B; p)$ is trivial. By Shapiro's Lemma this yields

$$H^{n+1}(G, \pi_*\pi^* B) = (0) .$$

However, we then have the exact sequence of G-modules

$$0 \to C \to \pi_*\pi^*B \xrightarrow{\mathfrak{N}} B \to 0$$

where $\mathfrak{N}$ is the (norm) map introduced in Proposition 8. The cohomology sequence then yields

$$\longrightarrow H^{n+1}(G, \pi_*\pi^*B; p) \longrightarrow H^{n+1}(G, B; p) \longrightarrow H^{n+2}(G, C; p) \longrightarrow \cdots$$

The left extreme vanishes by what we have just proved, and the right extreme vanishes as $scd_p G \leqq cd_p G + 1 = n + 1$. It follows that $H^{n+1}(G, B; p) = (0)$, and all is proved. Q.E.D.

PROPOSITION 19. *Let S be an open subgroup of* G, *and suppose* $cd_p G < \infty$. *Then* $cd_p S = cd_p G$.

Proof: Let G_p be a p-Sylow subgroup of G, then $cd_p(G_p) = cd(G_p) = cd_p G < \infty$; hence, $cd_p S = cd(S_p) \leqq cd(G_p)$. Moreover, S_p, being a p-subgroup of G, is contained in one of the p-Sylow subgroups of G, say the group G_p. Then $G_p \cap S = S_p$, so $(G_p : S_p) = (G_p : G_p \cap S) \leqq (G : S) < \infty$. All these remarks show that we may replace G by G_p and S by S_p for the proof. Let us do so, i.e., we now assume G is a p-group, S is a p-group and $(G:S) < \infty$.

If $n = \mathrm{cd}_p G$, Theorem 12 shows that all will follow if we can prove $H^n(S, Z/pZ) \neq (0)$. Since G acts trivially on Z/pZ, $\pi_{*_{S \to G}}(Z/pZ)$ is exactly the product of $(G:S)$ copies of Z/pZ, call it A. Shapiro's Lemma shows us that

$$H^n(G, A) \xrightarrow{\sim} H^n(S, Z/pZ) \quad ,$$

and so we must prove $H^n(G, A) \neq (0)$. However, A is a p-power G-module; hence, by Proposition 17, it possesses a maximal submodule A_1 with quotient Z/pZ.

$$0 \longrightarrow A_1 \longrightarrow A \longrightarrow Z/pZ \longrightarrow 0 .$$

If we apply cohomology, we get

$$\cdots \longrightarrow H^n(G, A) \longrightarrow H^n(G, Z/pZ) \longrightarrow H^{n+1}(G, A_1) \longrightarrow \cdots$$

Now, $\mathrm{cd}_p G = n$, so that $H^{n+1}(G, A_1) = (0)$ and $H^n(G, Z/pZ) \neq (0)$ by Theorem 12, again. It follows that $H^n(G, A) \neq (0)$, as required. Q.E.D.

COROLLARY 1. *If* G *is a* p*-group of cohomological dimension* $n < \infty$, *then for any finite* G*-module* A *of* p*-power order,* $H^n(G, A) \neq (0)$.

Proof: This is the content of the last few lines of the proof of Proposition 19.

COROLLARY 2. *If* $0 < \mathrm{cd}_p G < \infty$, *then* p^∞ *divides* $|G|$. *In particular, a finite group with non-zero* p*-Sylow subgroup has infinite* p*-cohomological dimension.*

Proof: We may assume G is a p-group, as is clear. If $p^\infty \nmid |G|$, then G is a finite group. Were $\mathrm{cd}_p(G) < \infty$, Proposition 19 would imply $0 = \mathrm{cd}_p\{1\} = \mathrm{cd}_p(G)$, a contradiction. Q.E.D.

If follows from Corollary 2 above, that the cohomological dimension theory for finite groups is empty. We still have not proven that there exist groups of every non-trivial cohomological dimension. Actually this is fairly hard, and we shall deal with it in the next chapter. Now we wish to prove a very important technical result which allows one to prove that certain groups have non-trivial dimension—and is useful in many contexts.

THEOREM 13. (Tower Theorem). *Let* N *be a closed normal subgroup of the profinite group* G. *Then*

$$\operatorname{cd}_p G \leqq \operatorname{cd}_p N + \operatorname{cd}_p G/N \quad . \tag{26}$$

If $\operatorname{cd}_p N = n < \infty$, *and* $\operatorname{cd}_p G/N = m < \infty$, *then we have an isomorphism*

$$H^{n+m}(G, A; p) \longrightarrow H^m(G/N, H^n(N, A); p) \quad ;$$

consequently, in this case if N *is a* p*-group and* $H^n(N, Z/pZ)$ *is finite, then* $\operatorname{cd}_p G = \operatorname{cd}_p N + \operatorname{cd}_p G/N$. *Moreover, the equality will also be valid if* N *is contained in the center of* G.

Proof: Consider the Hochschild-Serre Spectral Sequence

$$H^p(G/N, H^q(N, A)) \underset{p}{\Longrightarrow} H^*(G, A) \quad .$$

The groups $E_\infty^{p,q}$ are the composition factors for $H^{p+q}(G, A)$, and if $p + q > m + n$, then at each p, q we have either $p > m$ or $q > n$. If follows that $E_2^{p,q} = (0)$ for $p + q > m + n$; hence, its homology $E_3^{p,q}$ also vanishes, and so on. Therefore $E_\infty^{p,q} = (0)$ for $p + q > m + n$, and all composition factors for $H^{p+q}(G, A)$ (or its p-primary part) vanish when $p + q > m + n$. Equation (26) now follows immediately.

So now assume $\operatorname{cd}_p N = n < \infty$, $\operatorname{cd}_p G/N = m < \infty$. Let $(G/N)_p$ be a p-Sylow subgroup of G/N, and let G′ be its inverse image in G. The group G′ is a closed subgroup of G, so that $\operatorname{cd}_p G' \leqq \operatorname{cd}_p G$. We may therefore replace G by G′; hence, we may and do assume that G/N is a p-group. But then, $H^n(N, Z/pZ)$ is a finite p-group; hence, Corollary 1 of Proposition 19 implies that

$$H^m(G/N, H^n(N, Z/pZ); p) \neq (0) \quad ,$$

and all we must prove is the isomorphism of groups stated in the theorem.

Look at $E_2^{m,n}(p) = H^m(G/N, H^n(N, Z/pZ); p)$. We have

$$d_2^{m,n}: E_2^{m,n} \to E_2^{m+2,n-1} = (0); \text{ hence, } Z_2^{m,n} = E_2^{m,n} .$$

Also, $B_2^{m,n} = \mathrm{Im}(d_2^{m-2,n+1}) = (0)$, as $E_2^{m-2,n+1}(p) = (0)$. It follows that $E_3^{m,n} = E_2^{m,n}$. The same reasoning shows that $E_2^{m,n} = \cdots = E_\infty^{m,n}$ Moreover, one sees by exactly the same arguments that $E_\infty^{m,n}$ is the only non-zero term on the line $p + q = m + n$. Therefore we do indeed have the isomorphism

$$H^{m+n}(G, A; p) \xrightarrow{\sim} H^m(G/N, H^n(N, A); p) \quad .$$

The last case is when N is contained in the center of G. As before, we may assume G/N is a p-group. Now N is abelian, so its Pontrjagin dual is the direct sum of its p-primary components. It follows that N is the product of its p-Sylow subgroups; so we may write $N = N' \times N_p$ where N_p is the p-Sylow group, and N' the prime-to-p part of N. Then $H^q(N', Z/pZ)$ vanishes for all $q > 0$, so we deduce

$$H^n(N, Z/pZ) = H^n(N_p, Z/pZ) \neq (0) \quad .$$

(Use the Hochschild-Serre sequence for the group extension $0 \to N_p \to N \to N' \to 0$.) This group is a vector space over Z/pZ, and we may then write

$$H^n(N, Z/pZ) = (Z/pZ)^I$$

for some indexing set I. However, the action of G/N on $H^n(N, Z/pZ)$ is given by conjugation on the variables in N and action of G on Z/pZ. Both of these are trivial actions, the former because N lies in the center of G. Thus G/N acts trivially on $(Z/pZ)^I$, and we have proved

$$H^{n+m}(G, Z/pZ) = H^m(G/N, (Z/pZ)^I) \doteq H^m(G/N, Z/pZ)^I \neq (0) \; .$$

Q.E.D.

§2. The case of cohomological dimension one.

We shall study profinite groups G with $\mathrm{cd}_p G \leqq 1$ (essentially $\mathrm{cd}_p G = 1$ since the case $\mathrm{cd}_p G = 0$ has already been discussed). This will provide us with important insights into the structure of profinite groups as well as with our first example of groups of non-trivial cohomological dimension.

The first topic is a characterization of groups G with $\mathrm{cd}_p G \leqq 1$, and as this involves group extensions, we shall study these to begin with. By an *extension of a group* G *by a group* A, we mean an exact sequence

$$0 \to A \to E \to G \to 0\,, \tag{27}$$

so that A is the kernel and G is the quotient of E by A. From now on we assume A *is abelian,* G *is profinite,* E *is a topological group and all maps are continuous.*

To give a *continuous action of* G *on* A means merely to have a homomorphism $G \to \mathrm{Aut}(A)$ under which $G \times A \to A$ is a continuous function. Now observe that every extension of G by A gives rise to a continuous action of G on A. This is done as follows: The group E acts on A *via* inner automorphisms, *viz.*:

$$\sigma\colon\ a \mapsto \sigma a \sigma^{-1}\,,$$

this is continuous, and as A is abelian, it acts trivially on itself. It follows that the above action is really a $G = E/A$ action. So every extension (27) makes A a G-module. Given an extension (27), we shall refer to the induced action of G on A as the *type* of the extension (27).

THEOREM 14. *Let* A *be a finite abelian group, and let* G *be a profinite group. Then the (isomorphism classes of) extensions of* G *by* A *of given type are in* 1-1 *correspondence with the elements of* $H^2(G, A)$. *Corresponding to the zero element of* $H^2(G, A)$, *we have the split extension (of the given type), namely* $E = A \times G$ *with group multiplication*

$$\langle a, g\rangle \langle a', g'\rangle = \langle a + g\cdot a', gg'\rangle\ .$$

Proof: Let

$$(*) \qquad 0 \longrightarrow A \longrightarrow E \xrightarrow{\phi} G \longrightarrow 0$$

be a given extension. Since A is a finite group, the group E is profinite and Theorem 3 (the cross-section theorem) yields the existence of a continuous section, say i, for ϕ. Now consider the function on $G \times G$ to E defined by

$$f(\sigma,\tau) = i(\sigma)\, i(\tau)\, i(\sigma\tau)^{-1} \quad .$$

Clearly, $\phi(f(\sigma,\tau))$ vanishes; hence, $f \in C^2(G, A)$. Write A multiplicatively and consider

$$\delta f(\sigma,\tau,\rho) = \sigma f(\tau,\rho)\, f(\sigma\tau,\rho)^{-1} f(\sigma,\tau\rho)\, f(\sigma,\tau)^{-1} \quad .$$

Because A is abelian, and because $\sigma f(\tau,\rho)$ is $i(\sigma)\, f(\tau,\rho)\, i(\sigma)^{-1}$ by definition, we get

$$\begin{aligned}(\delta f)(\sigma,\tau,\rho) = {} & i(\sigma)\, i(\tau)\, i(\rho)\, i(\tau\rho)^{-1} i(\sigma)^{-1} i(\sigma)\, i(\tau\rho)\, i(\sigma\tau\rho)^{-1} i(\sigma\tau\rho) \\ & \cdot\, i(\rho)^{-1} i(\sigma\tau)^{-1} i(\sigma\tau)\, i(\tau)^{-1} i(\sigma)^{-1} = 1 \ ,\end{aligned}$$

and this shows that f is a 2-cocycle of G with coefficients in A. If $j(\sigma)$ is another cross-section for ϕ, let $k(\sigma) = i(\sigma)\, j(\sigma)^{-1}$, and let $g(\sigma,\tau)$ be the cocycle induced by the cross-section $j(\sigma)$ and our given extension (*). A simple, explicit calculation shows that

$$\delta k(\sigma,\tau) = f(\sigma,\tau)\, g(\sigma,\tau)^{-1} \ ,$$

so that f is cohomologous to g. It follows that the cohomology class, $\alpha = \theta(*)$, of the cocycle f is dependent only upon the extension (*).

Recall that two extensions (*) and (*)′

$$(*)' \qquad 0 \longrightarrow A \longrightarrow E' \longrightarrow G \longrightarrow 0$$

are isomorphic (or equivalent) iff there is a commutative diagram

$$\begin{array}{ccccccccc} 0 & \longrightarrow & A & \longrightarrow & E & \longrightarrow & G & \longrightarrow & 0 \\ & & \big\downarrow \text{id} & & \big\downarrow \Phi & & \big\downarrow \text{id} & & \\ 0 & \longrightarrow & A & \longrightarrow & E' & \longrightarrow & G & \longrightarrow & 0 \end{array}$$

in which Φ is an isomorphism. One checks very readily that two isomorphic extensions yield cohomologous cocycles (use as section for $G \to E'$ the composed map $\Phi \circ i$); this proves that θ maps isomorphism classes of extensions to elements of $H^2(G, A)$.

To prove θ is the desired 1-1 mapping, we shall construct an explicit inverse mapping, ω. Let $\beta \in H^2(G, A)$ and let $g(\sigma, \tau)$ be any cocycle representing β. Let $E = A \times G$ as sets, and define multiplication in E *via*

$$\langle a, \sigma\rangle\langle b, \tau\rangle = \langle a(\sigma b)\, g(\sigma, \tau),\ \sigma\tau\rangle .$$

The reader should check that the *associativity of this multiplication is a direct consequence of the cocycle identity* $\delta g(\sigma, \tau, \rho) = 1$. Moreover, it is clear that E is a group and that

$$0 \to A \to E \to G \to 0$$

is exact. If $h(\sigma, \tau)$ is another cocycle, cohomologous to g, form the set $A \times G$ as before and call it E' when endowed with the multiplication which the cocycle h induces (as above for g). Since $g \sim h$, there exists a 1-cochain $k(\sigma)$ such that

$$g(\sigma, \tau)\, h(\sigma, \tau)^{-1} = \sigma k(\tau)\, k(\sigma\tau)^{-1} k(\sigma) .$$

Consider the mapping $\Phi: E \to E'$ given by $\Phi(\langle a, \sigma\rangle) = \langle ak(\sigma), \sigma\rangle$. One checks easily that Φ is an isomorphism (observe that $k(1) = 1$) which makes E and E' isomorphic extensions. Therefore, the mapping $\omega : \beta \mapsto$ class of E is well-defined and is obviously the inverse of θ.

The last assertion of the theorem concerning split extensions now follows immediately, and we are done. Q.E.D.

Given a diagram

$$\begin{array}{ccccccccc} & & & & & & G & & \\ & & & & & & \downarrow{\scriptstyle f} & & \\ 0 & \longrightarrow & A & \longrightarrow & E & \underset{\phi}{\longrightarrow} & W & \longrightarrow & 0 \end{array}$$

in which the row is exact, one can form an extension of G *via* f which is known as the *pull-back* of the extension $0 \to A \to E \to W \to 0$ (or the *inverse image* of this extension) *by* f. We do this as follows: E_f shall be the *fibred product of* E *and* G *over* W, that is

$$E_f = \{\langle e, \sigma\rangle \in E \times G \mid \phi(e) = f(\sigma)\} = E \times_W G \ .$$

There is a natural map $E_f \to G$, call it π, whose kernel is precisely A. We obtain the commutative diagram,

$$\begin{array}{ccccccccc} 0 & \longrightarrow & A & \longrightarrow & E_f & \xrightarrow{\pi} & G & \longrightarrow & 0 \\ & & \Big\downarrow{\scriptstyle id} & & \Big\downarrow & & \Big\downarrow{\scriptstyle f} & & \\ 0 & \longrightarrow & A & \longrightarrow & E & \xrightarrow[\phi]{} & W & \longrightarrow & 0 \ ; \end{array} \tag{28}$$

and if (E) denotes the extension of the bottom exact sequence, we let $f^*((E))$ denote the top exact sequence.

Let $0 \to P \to E \xrightarrow{\phi} W \to 0$ be an extension, (E), of profinite groups. We shall say that the profinite group G has the *lifting property for the extension* (E) if and only if for every homomorphism $f: G \to W$, there is a "lifting" $f': G \to E$ so that $f = \phi \circ f'$.

PROPOSITION 20. *The profinite group* G *has the lifting property for the extension* (E) *if and only if the inverse image,* $f^*((E))$, *of* (E) *splits for every* $f: G \to W$.

Proof: Assume the lifting property (refer to (28)), and define $i: G \to E_f$ by $i(\sigma) = \langle f'(\sigma), \sigma\rangle$. One checks immediately that i is a splitting for $f^*((E))$. Conversely, assume $f^*((E))$ splits, say i is a splitting homomorphism: $G \to E_f$. The composition $G \to E_f \to E$ is then the required lifting f' of f. Q.E.D.

We say a profinite group G is p-*extensive* if and only if it has the lifting property for all extensions in which P is a finite, abelian p-group. Here is the advertised connection between extensions and cohomological dimension.

THEOREM 15. *For a profinite group* G, *the following statements are equivalent.*

(a) $cd_p G \leqq 1$

(b) G *is* p-*extensive*

(c) *Every extension of* G *by a finite, abelian* p-*group splits*

(d) G *has the lifting property for all extensions* (E) *in which* P *is a profinite* p-*group*

(e) *Every extension of* G *by a profinite* p-*group is trivial.*

Proof: (a) $\Longrightarrow$ (c). By Theorem 14 extensions such as those described in (c) are classified by $H^2(G, P)$ which vanishes by (a). Hence, (c) follows. In the same way, arguing backwards, we see that (c) $\Longrightarrow$ (a). The conclusion (b) $\Longleftrightarrow$ (c) is immediate from Proposition 20, as is the conclusion (d) $\Longleftrightarrow$ (e). Clearly (e) $\Longrightarrow$ (c), so all we need to prove is (c) $\Longrightarrow$ (e).

We consider an extension

(E) $$0 \to P \to E \to G \to 0$$

in which P is a profinite p-group. Let $\mathcal{S}$ be the set of all pairs $\langle P_\alpha, \phi_\alpha \rangle$ where P_α is closed in P, normal in E, and ϕ_α is a homomorphism $G \to E/P_\alpha$ lifting the identity map $G \to G$. Clearly, $\langle P, 1 \rangle \in \mathcal{S}$, so $\mathcal{S}$ is non-empty and is partially ordered in the usual way. By Zorn's Lemma, pick a maximal pair $\langle P_0, \phi_0 \rangle$ from $\mathcal{S}$. Then P_0 is closed in P, normal in E, and $\phi_0 : G \to E/P_0$ lifts $G \to G$. We claim $P_0 = \{1\}$. If not, because P_0 is a p-group, there would exist a *non-trivial* homomorphism $\chi : P_0 \to Z/pZ$. Now E/P_0 acts on $H^1(P_0, Z/pZ)$ in the Hochschild-Serre manner, namely $\chi^\sigma(\tau) = \sigma \cdot \chi(\sigma\tau\sigma^{-1}) = \chi(\sigma\tau\sigma^{-1})$ and as this operation is continuous, there exist only finitely many conjugates, say $\chi_1, \ldots, \chi_r$, for χ. (Say $\chi = \chi_1$.) Let $\tilde{P} = \cap_i (\ker \chi_i)$, then $\tilde{P}$ is closed in P_0; hence closed in P. Moreover, $\tilde{P}$ is normal in E by construction, and the index $(P_0 : \tilde{P})$ is finite ($\leqq p^r$). Now as $P_0 \neq Z/pZ$ (else assumption (c) would allow an instant extension of ϕ_0, contradicting the maximality of $\langle P_0, \phi_0 \rangle$), we deduce that $\tilde{P}$ is a *proper* subgroup of P_0. We are now going to show that we can extend ϕ_0 to a lifting $\tilde{\phi} : G \to E/\tilde{P}$; i.e., we can embed the diagram (without the dotted arrow)

$$\begin{array}{ccccccccc} 0 & \longrightarrow & P_0/\tilde{P} & \longrightarrow & E/\tilde{P} & \longrightarrow & E/P_0 & \longrightarrow & 0 \\ & & & & & \tilde{\phi}\nwarrow & \uparrow \phi_0 & & \\ & & & & & & G & & \end{array}$$

into the diagram with the dotted arrow. Since $P_0/\tilde{P}$ is finite, we may argue by induction on the length of a composition series; hence, we may and do assume $P_0/\tilde{P}$ is simple. But a simple p-group is $\mathbf{Z}/p\mathbf{Z}$, and in this case hypothesis (c) yields the required map $\tilde{\phi}$ at once. This contradicts the maximality of $\langle P_0, \phi_0 \rangle$, and it finally proves $P_0 = \{1\}$. Q.E.D.

PROPOSITION 21. *Let* X *be a set, and let* $F_p(X)$ *be the free profinite* p*-group on* X *(cf. example (4)). The homomorphisms of* $F_p(X)$ *into a discrete (in particular, finite)* p*-group* P *are in* 1-1 *correspondence with the set-theoretic maps* $X \to P$ *such that almost all* $x \in X$ *are mapped to* $0 \in P$. *Consequently,* $F_p(X)$ *is* p*-extensive, and thus* $\mathrm{cd}_p F_p(X) \leqq 1$.

Proof: If $\xi \in \mathrm{Hom}_C(F_p(X), P)$, then $\xi^{-1}(0)$ is open and normal in $F_p(X)$, call it V. If $F(X)$ is the ordinary free group on X, then there is the canonical injection $F(X) \to F_p(X)$, and so $V \cap F(X) = U$ is a normal subgroup of $F(X)$ of p-power index. Hence, ξ gives us a homomorphism $F(X)/U \to P$, and by composition a homomorphism $F(X) \to P$. By the continuity of ξ, almost all $x \in X$ map to zero under the set-theoretic map $X \to P$ induced by ξ. The converse is obtained by reading backwards in the above argument.

If the diagram

$$\begin{array}{ccccccccc} 0 & \longrightarrow & P & \longrightarrow & E & \longrightarrow & W & \longrightarrow & 0 \\ & & & & & \nwarrow & \uparrow \xi & & \\ & & & & & & F_p(X) & & \end{array}$$

is given where P is a finite p-group, we may assume E is a p-group, for we need only take a p-Sylow subgroup of E which contains P and map $F_p(X)$

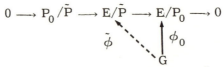

to this subgroup. However, ξ corresponds to a map $X \to W$ almost all $x \in X$ mapping to zero in W. Any set-theoretic lifting of this map to E, preserving zero, will induce the desired lifting shown in the dotted arrow. (Here, of course, for the p-extensivity we need only treat the case of finite E.)

Q.E.D.

COROLLARY 1. *If* X *is a set and* $|X|$ *denotes its cardinal number, then* $H^1(F_p(X), Z/pZ)$ *is the vector space of dimension* $|X|$ *over* Z/pZ. *Consequently,* $cd_p(F_p(X)) = cd(F_p(X)) = 1$ *for every non-empty* X.

COROLLARY 2. $F_p(X)$ *is isomorphic to* $F_p(Y)$ *if and only if* $|X| = |Y|$. *Thus for each* $|X|$, *there exists one and only one free p-group whose generating set is equipotent with* $|X|$. *This cardinal number is the unique invariant of* $F_p(X)$.

COROLLARY 3. $\hat{Z}$ (cf. example 4) *has cohomological dimension* 1.

Proof: $\hat{Z}$ has $Z_p = F_p(X)$ (where X is the one point set) as p-Sylow subgroup for each p. Q.E.D.

§3. The Structure of Profinite p-groups; Šafarevič-Golod's Theorem

For all of this section, G will denote a profinite p-group.

LEMMA 2. *Let* S *be a proper closed subgroup of* G. *Then there exists a non-zero element* $\chi \in H^1(G, Z/pZ)$ *such that* $\chi | S \equiv 0$.

Proof: Since $S \neq G$, there exists an open normal subgroup U such that the image of S in G/U is not all of G/U. Call the image S_0. Since G/U is a finite p-group, it is solvable; so there exists a normal subgroup N of G/U of index p for which $S_0 \subseteq N$. Thus, the composed homomorphism $G \to G/U \to Z/pZ = (G/U)/N$ is the required χ. Q.E.D.

Now, Lemma 2 prompts us to define the *Frattini subgroup* of G, denoted G^*, by the equation

(29) $$G^* = \bigcap \{\ker \chi \mid \chi \in H^1(G, Z/pZ)\}\,.$$

Clearly, G^* is a closed, normal subgroup of G.

PROPOSITION 22. *The Frattini subgroup,* G^*, *of the profinite* p-*group* G *has the following properties:*

(a) G/G^* *is Pontrjagin dual to* $H^1(G, Z/pZ)$;

(b) *If* S *is a closed subgroup of* G, *then* $S = G$ *if and only if* $SG^* = G$;

(c) *Any homomorphism of profinite* p-*groups* $G \xrightarrow{\phi} \tilde{G}$ *induces a corresponding homomorphism* $\phi_*: G/G^* \to \tilde{G}/\tilde{G}^*$. *The homomorphism* ϕ *is surjective iff* ϕ_* *is surjective.*

(d) $G^* = G^p \cdot \overline{[G,G]}$.

Proof: (a) $H^1(G, Z/pZ)$ has the discrete topology, so its dual $H^1(G, Z/pZ)^D$ is compact; in fact, its dual has the product topology. If $\sigma \in G$, let $\tilde{\sigma}(\chi) = \chi(\sigma)$, then $\sigma \mapsto \tilde{\sigma}$ is a continuous homomorphism from G to $H^1(G, Z/pZ)^D$. Clearly, the kernel of this homomorphism is G^*; so the sequence

$$0 \longrightarrow G^* \longrightarrow G \xrightarrow{\theta} H^1(G, Z/pZ)^D$$

is exact. If $\lambda \in H^1(G, Z/pZ)^D$, then $\lambda \in H^1(G^{ab}, Z/pZ)^D$; and by Pontrjagin duality this shows that $\lambda \in G^{ab}$. Hence, there exists an element $\lambda_0 \in G$ inducing λ, i.e., $\theta(\lambda_0) = \lambda$, and (a) follows.

(b) If $S = G$, then $SG^* = G$. If $S \neq G$, then Lemma 2 provides us with an element $\chi \in H^1(G, Z/pZ)$ which vanishes on S yet not on all of G. Since χ kills G^* by definition, we deduce $\chi \mid SG^* \equiv 0$. It follows that $G \neq SG^*$.

(c) Let $\phi: G \to \tilde{G}$ be given. Then there is the induced homomorphism $H^1(\tilde{G}, Z/pZ) \to H^1(G, Z/pZ)$; by duality, and part (a) we get $\phi_*: G/G^* \to \tilde{G}/\tilde{G}^*$. If $G \xrightarrow[\phi]{} \tilde{G} \longrightarrow 0$ is exact, then inflation from $H^1(\tilde{G}, Z/pZ) \to H^1(G, Z/pZ)$ is injective; so by duality, ϕ_* is surjective.

Suppose, conversely, ϕ is not surjective. Let $\tilde{S} = \phi(G)$, then $\tilde{S}$ is a proper closed subgroup of $\tilde{G}$, so by Lemma 2, we can find a $\tilde{\chi}$ in $H^1(\tilde{G}, Z/pZ)$ which vanishes on $\tilde{S}$ but not on $\tilde{G}$. Clearly the homomorphism $H^1(\tilde{G}, Z/pZ) \to H^1(G, Z/pZ)$ kills $\tilde{\chi}$; so its dual ϕ_* is not surjective.

(d) It follows from (a) that $G/G^* \xrightarrow{\sim} G^{ab}/(G^{ab})^*$ (where G^{ab} = maximal abelian factor group of G); so we may assume at once that $G = G^{ab}$. Clearly, if $\xi \in G^p$, then ξ is killed by every $\chi \in H^1(G, Z/pZ)$, i.e., $G^p \subseteq G^*$. Conversely, consider $(G/G^p)^D$. We have $(G/G^p)^D = H^1(G/G^p, Z/pZ)$, and if $\xi \in G^*$ with non-zero image in G/G^p, say $\bar{\xi}$, then there is a $\bar{\chi}$ in $(G/G^p)^D$ such that $\bar{\chi}(\bar{\xi}) \neq 0$. But then $\bar{\chi}$ comes from some $\chi \in H^1(G, Z/pZ)$, and $\chi(\xi) \neq 0$ – so $\xi \notin G^*$, a contradiction whence we deduce $\bar{\xi} = 0$; that is, $G^* \subseteq G^p$. Q.E.D.

Example 8. Let $G = F_p(X)$. Then $G/G^* \xrightarrow{\sim} \Pi_X Z/pZ$.

The Frattini subgroup gives us a very good hold on the crude structure of profinite p-groups. We shall make this explicit in the next proposition which the reader may think of as the profinite analog of the Burnside basis theorem.

PROPOSITION 23. *Let* G *be a profinite* p-*group, and let* X *be a set. If* $\theta : H^1(G, Z/pZ) \to \Pi_X Z/pZ$ *is a vector space homomorphism, then there is a corresponding group homomorphism,* $\pi : F_p(X) \to G$ *such that* $H^1(\pi, Z/pZ)$ *is* θ. *Moreover, if* θ *is injective then* π *is surjective; and if* θ *is bijective and* $cd_p G \leq 1$, *then* π *is an isomorphism.*

Proof: The Pontrjagin dual of θ, say θ^D, takes $\Pi_X Z/pZ$ to $H^1(G, Z/pZ)^D = G/G^*$. Thus, by example 8 the map θ^D takes $F_p(X)/F_p(X)^*$ to G/G^*. Consequently, we have the diagram

$$\begin{array}{ccccccccc} 0 & \longrightarrow & G^* & \longrightarrow & G & \longrightarrow & G/G^* & \longrightarrow & 0 \; , \\ & & & & & {\nwarrow}^{\pi} & \uparrow & & \\ & & & & & & F_p(X) & & \end{array}$$

and this yields π, as shown, because $cd_p F_p(X) \leqq 1$. Suppose θ is injective, then $F_p(X) \to F_p(X)/F_p(X)^* \to G/G^*$ is surjective. If S is the image of π, then $SG^* = G$, so by (b) π is surjective. Lastly, suppose θ is bijective, then $\pi: F_p(X) \to G$ is surjective as we have just shown. But $cd_p G \leqq 1$, so the group extension

$$0 \longrightarrow K \longrightarrow F_p(X) \xrightarrow{\pi} G \longrightarrow 0$$

splits, say ρ is the section of π. Since π is surjective, ρ is injective (we know $\pi \circ \rho = 1$). Now $H^1(1) = H^1(\pi \circ \rho) = H^1(\rho) \circ H^1(\pi) = H^1(\rho) \circ \theta$. As θ is *bijective*, so is $H^1(\rho)$; it follows from this that ρ is surjective. And now the bijectivity of ρ, and $\pi \circ \rho = 1$ prove that ρ is the inverse map to π. Q.E.D.

COROLLARY 1. *In order that a profinite* p-*group* G *be a quotient of* $F_p(X)$ *for the set* X, *it is necessary and sufficient that* $H^1(G, \mathbb{Z}/p\mathbb{Z})$ *have a basis whose cardinal number is less than or equal to* $|X|$. *Consequently, every profinite* p-*group is a homomorphic image of a free profinite* p-*group.*

COROLLARY 2. *If* G *is a profinite* p-*group, then* $cd_p G \leqq 1$ *if and only if* $G = F_p(X)$ *for some set* X. *More precisely, if and only if* $G = F_p(B)$ *where* B *is a basis for* $H^1(G, \mathbb{Z}/p\mathbb{Z})$.

COROLLARY 3. *Every closed subgroup of a free profinite* p-*group is free.*

COROLLARY 4. *The strong* p-*completion (cf. example* (3P)*),* $\hat{\hat{F}}(X)$, *of the free group,* F(X), *is a free profinite group. Moreover, if* X *is finite this completion is* $F_p(X)$; *while if* X *is infinite, the completion is* $F_p(2^X)$.

Proof: $\hat{\hat{F}}(X)$ has the lifting property for all extensions with finite abelian kernel because the homomorphisms of $\hat{\hat{F}}(X)$ into a finite abelian p-group, P, are in 1-1 correspondence with the set-theoretic maps from X to P. Thus, Corollary 2 applies, and we have

$$\dim H^1(\hat{\hat{F}}(X), Z/pZ) = \begin{cases} |X| & \text{if } X \text{ is finite} \\ |2^X| & \text{if } X \text{ is infinite} \end{cases} .$$

Q.E.D.

Let G be a profinite group, and let $\sigma_1, \ldots, \sigma_n$ be elements of G. Let us say that $\sigma_1, \ldots, \sigma_n$ *generate* G *topologically* if and only if the closure of the ordinary group generated by $\sigma_1, \ldots, \sigma_n$ is all of G. It is the same to require that for each open normal U of G, the images $\bar{\sigma}_1, \ldots, \bar{\sigma}_n$ of the σ's generate G/U.

The connection between the notion of topological generation and free p-groups is the following very simple proposition whose proof will be left as an exercise.

PROPOSITION 24. *Let* G *be a profinite* p-*group and let* $\sigma_1, \ldots, \sigma_n$ *be a set of elements of* G. *The following statements are equivalent:*

(a) $\sigma_1, \ldots, \sigma_n$ *generate* G *topologically;*

(b) *The homomorphism* $F_p(X) \to G$ *induced by* $\sigma_1, \ldots, \sigma_n$ (where $|X| = n$) *is surjective;*

(c) *The elements* $\bar{\sigma}_1, \ldots, \bar{\sigma}_n$ *of* G/G^* *generate* G/G^* *topologically;*

(d) *Every* $\chi \in H^1(G, Z/pZ)$ *which vanishes on the* σ_j $(j = 1, 2, \ldots, n)$ *is identically zero.*

COROLLARY. *The minimal number of generators of* G *is the dimension of* $H^1(G, Z/pZ)$; hence, G *is finitely generated if and only if* G^* *is open in* G.

Suppose F is a profinite p-group and N is a closed normal subgroup of F. Let $\tau_1, \ldots, \tau_n$ be elements of N. We shall say that $\tau_1, \ldots, \tau_n$ *generate* N *as normal subgroup* if and only if the τ's *and their conjugates under* F generate N as profinite group. To prove the next step in our march toward a better understanding of presentations of profinite groups, we need

LEMMA 3. *Let* G *be a profinite* p*-group and let* A *be a* p*-torsion* G*-module. If* $A^G = (0)$, *then* $A = (0)$.

Proof: Write A as the limit of finitely generated submodules A_α, then $A^G = \text{dir lim } A_\alpha^G$; so we may and do assume A is finitely generated. In this case A has a composition series with all its factors Z/pZ, each having trivial action. Induction on the length of such a series now completes our proof. Q.E.D.

PROPOSITION 25. *In order that* $\tau_1, \ldots, \tau_n$ *generate* N *as normal subgroup of the profinite* p*-group* F, *it is necessary and sufficient that every* $\chi \in H^1(N, Z/pZ)^{F/N}$ *which vanishes on all the* τ_j *vanishes on all of* N.

Proof: If the τ's generate N in the sense of normality, then $\{\sigma\tau_j\sigma^{-1} \mid \sigma \in F\}$ generates N in the topological sense. Hence, Proposition 24 shows that all χ in $H^1(N, Z/pZ)$ which vanish on $\sigma\tau_j\sigma^{-1}$ for all $\sigma \in F$ will vanish on all of N. But if χ is fixed by F/N, then $\chi(\sigma\tau_j\sigma^{-1}) = \chi(\tau_j)$ —so if these χ vanish on the τ_j they will vanish on all of N.

For the converse, let τ's be given with the vanishing property above, and let N′ be the smallest, closed, normal subgroup of F containing $\tau_1, \ldots, \tau_n$. Since $N' \hookrightarrow N$ we have the restriction mapping

$$H^1(N, Z/pZ) \to H^1(N', Z/pZ)$$

and this respects the operation of F on these groups. So, we have a homomorphism

$$\overline{\text{res}}: H^1(N, Z/pZ)^F \to H^1(N', Z/pZ)^F .$$

If $\xi \in \ker(\overline{\text{res}}) = (\ker(\text{res}))^F$, then ξ restricted to N′ vanishes, so ξ is an invariant character vanishing on the τ_j and all their conjugates. Thus, by hypothesis, ξ vanishes on all of N, and we deduce $(\ker(\text{res}))^F = (0)$. Now, Lemma 3 shows that res is an injection, so by duality

$N'/N'^* \to N/N^*$ is surjective. But then by part (c) of Proposition 22, we conclude that our injection $N' \to N$ is really surjective, too; and we are done. Q.E.D.

COROLLARY. *The normal subgroup* N *is generated as normal subgroup of* F *by* n *elements if and only if the dimension of* $H^1(N, Z/pZ)^{F/N}$ *is less than or equal to* n.

These arguments show that the dimension of $H^1(N, Z/pZ)^{F/N}$ (the so-called *rank of* N *as normal subgroup of* F), call it $r_F(N)$, is the minimal number of elements $\tau_1, \ldots, \tau_r$ which can generate N as normal subgroup of F.

As standard notation, let $h_i(G) = \dim H^i(G, Z/pZ)$ for all $i > 0$, and all profinite p-groups G.

PROPOSITION 26. *If* G *is a profinite* p*-group which is finitely generated* (i.e., $h_1(G) < \infty$), *and if*

$$0 \longrightarrow N \longrightarrow F_p(X) \longrightarrow G \longrightarrow 0$$

is exact for some finite set X *of cardinality* n, *then* $r_F(N)$ *is finite if and only if* $h_2(G)$ *is finite* (i.e., *if and only if* $H^2(G, Z/pZ)$ *is a finite group*). *Moreover, when this is the case, we have the equation*

$$r_F(N) = n - h_1 + h_2 \quad .$$

Proof: The inflation-restriction sequence

$$0 \to H^1(G, Z/pZ) \to H^1(F_p(X), Z/pZ) \to H^1(N, Z/pZ)^{F/N} \to H^2(G, Z/pZ) \to 0$$

is an exact sequence of vector spaces over Z/pZ. The alternating sum of the dimensions is then zero and everything follows immediately from this.

Q.E.D.

COROLLARY. *Let* G *be a finitely generated profinite* p*-group, and let* B *be a basis for* $H^1(G, Z/pZ)$. *Then in the canonical presentation*

$$0 \to N \to F_p(B) \to G \to 0 \ ,$$

the integer $r_F(N)$ *is exactly* $h_2(G)$. *Therefore,* $h_2(G)$ *is the number of relations among any minimal set of generators defining* G.

If G is a *finite* p-group it certainly is a profinite p-group; so the preceeding analysis applies to such a G. The theorem of Šafarevič-Golod deals with the integer $h_2(G) - h_1(G) = \underline{\sigma}(G)$ for finite p-groups. It has extremely important applications in class field theory (which we give below) and in the theory of algebras [SG]. Let us call the integer $\underline{\sigma}(G)$ the *Šafarevič number* of G. The first result is the non-negativity of $\underline{\sigma}(G)$.

PROPOSITION 27. *For every finite* p*-group,* G, *the Šafarevič number is non-negative. Moreover,* $\underline{\sigma}(G)$ *is the "rank" of* $H^3(G, Z)$, *that is, the number of cyclic factors of* $H^3(G, Z)$.

Proof: We need prove only the last statement. Consider the exact sequence $0 \to Z \xrightarrow[p]{} Z \to Z/pZ \to 0$, and apply cohomology:

$$0 \to H^1(G, Z/pZ) \to H^2(G, Z) \xrightarrow{p} H^2(G, Z) \to H^2(G, Z/pZ) \to H^3(G, Z)_p \to 0 \ .$$

Here, $H^3(G, Z)_p$ is the kernel of the map $H^3(G, Z) \xrightarrow{p} H^3(G, Z)$. Now $H^2(G, Z)$ is the dual of G^{ab} (cf. §3 of Chapter II); hence, is a finite group, and from the sequence $H^3(G, Z)_p$ is also finite. It follows that the alternating product of the orders of these groups is 1,

$$\frac{p^{h_1(G)} \cdot |H^2(G, Z)| \cdot p^t}{|H^2(G, Z)|\, p^{h_2(G)}} = 1 \ .$$

Here, $t = \dim H^3(G, Z)_p$ = rank of $H^3(G, Z)$. Upon taking logarithms to the base p in the above product, we deduce

$$\underline{\sigma}(G) = h_2(G) - h_1(G) = t \geqq 0 \ . \qquad \text{Q.E.D.}$$

The proof of Šafarevič-Golod's Theorem depends upon some material in the homology theory of finite groups.* We shall give a short sketch (without proofs) of this theory and the relevant facts; for complete details the reader should consult [CE, SCL].

If G is a finite group, a G-module A is the same as a module over the group ring $Z[G]$. Hence, the category of G-modules possesses enough projective as well as injective objects. Let IA denote the submodule of A generated by the elements $(\sigma - 1)a$ as a ranges over A and σ ranges over G. We consider the functor $A \rightsquigarrow A/IA = H_0(G, A)$; it is right-exact from the category of G-modules, $\mathfrak{A}(G)$, to the category of abelian groups. Its left-derived functors, $L_n H_0(G, A)$, are denoted $H_n(G, A)$, and are the *homology groups of* G *with coefficients in* A. One may check that

$$H_0(G, A) = Z \otimes_{Z[G]} A \ ;$$

hence, it follows that

$$H_n(G, A) = \mathrm{Tor}_n^{Z[G]}(Z, A) \ .$$

There is, in fact, an explicit construction of these groups as homology groups of a certain "standard" complex (dual to the one we described for the cohomology theory of Chapter II). It goes as follows: For each $n \geqq 0$, let $C_n(G, A)$ — *the group of* n-*chains of* G *with coefficients in* A — be

$$C_n(G, A) = S_n(G) \otimes_{Z[G]} A$$

where $S_n(G)$ is the free abelian group on the cartesian product of G with itself $n+1$ times. This group is a $Z[G]$-module under the operation

$$\sigma(\sigma_0, \ldots, \sigma_n) = (\sigma\sigma_0, \ldots, \sigma\sigma_n) \ .$$

The boundary mapping $\partial_n : C_n(G, A) \to C_{n-1}(G, A)$ is given by the map $\partial_n : S_n(G) \to S_{n-1}(G)$ defined in turn by the equation

$$\partial_n(\sigma_0, \ldots, \sigma_n) = \sum_{i=0}^{n} (-1)^i (\sigma_0, \ldots, \hat{\sigma}_i, \sigma_{i+1}, \ldots, \sigma_n)$$

* The proof we give of the Šafarevič-Golod Theorem is due to Roquette, Vinberg, and Gaschutz, see [R].

where $(\sigma_0, \ldots, \hat{\sigma}_i, \ldots, \sigma_n)$ signifies the n-tuple obtained by omitting σ_i. Thus $\partial_n : C_n(G, A) \to C_{n-1}(G, A)$ is given by the formula:

$$\partial_n((\sigma_0, \ldots, \sigma_n) \otimes a) = \sum_{i=0}^{n} (-1)^i (\sigma_0, \ldots, \hat{\sigma}_i, \ldots, \sigma_n) \otimes a \ .$$

(This complex differs only in notation from the one described in Chapter II, it is *homogeneous*. Transition to the complex of Chapter II is effected by setting $[\sigma_1, \ldots, \sigma_n] = (1, \sigma_1, \sigma_1\sigma_2, \ldots, \sigma_1\sigma_2 \cdot \ldots \cdot \sigma_n)$, and by taking $\mathrm{Hom}_{Z[G]}(S_n(G), A)$ rather than the tensor product.)

Of course, one has the n-*cycles* $Z_n(G, A) = \ker \partial_n$, and the n-*boundaries* $B_n(G, A) = \mathrm{Im}\, \partial_{n+1}$, so that

$$H_n(G, A) = Z_n(G, A)/B_n(G, A)$$

is the explicit description of the homology of G.

Remark. All this may be done without assuming G is a finite group— merely a discrete group. Morevoer, proofs of the above statements may be constructed by appropriately dualizing the notion of a δ-functor so as to obtain a ∂-functor, and then proceeding just as in Chapter II, §§1 and 2.

The crucial point in all of this is the fact (discovered by Tate [TC]) that, when G is FINITE, one can splice together the homology of G and its cohomology, and by changing some numbers have a "cohomology theory" running from $-\infty$ to ∞. The core of Tate's argument is that when G is finite one has the *norm map* $\mathfrak{N} : A \to A$ given by

$$\mathfrak{N}a = \sum_{\sigma \in G} \sigma a \ .$$

Clearly, the image $\mathfrak{N}A$ of A under $\mathfrak{N}$ is fixed by all of G, so that $A^G \supseteq \mathfrak{N}A$; and equally clearly every element of IA is killed by $\mathfrak{N}$, $IA \subseteq A_{\mathfrak{N}} = \ker(A \xrightarrow{\mathfrak{N}} A)$. This led Tate to alter two groups and renumber the others as follows:

(30) $$\tilde{H}^0(G, A) = A^G/\mathfrak{N}A \ ,$$

(31) $$\tilde{H}^{-1}(G, A) = A_{\mathfrak{N}}/IA$$

(32) $$\tilde{H}^{-n}(G, A) = H_{n-1}(G, A) \quad \text{for } n \geqq 2$$

(33) $$\tilde{H}^{n}(G, A) = H^{n}(G, A) \quad \text{for } n > 0 \ .$$

These groups, so altered and renumbered, will be referred to as the *reduced cohomology groups of* G *with coefficients in* A. One now has

THEOREM 16. (Tate). *If* G *is a finite group, and if*

$$0 \to A' \to A \to A'' \to 0$$

is an exact sequence of G*-modules, then the sequence*

$$\cdots \to \tilde{H}^{-n}(G, A'') \to \tilde{H}^{-n+1}(G, A') \to \tilde{H}^{-n+1}(G, A) \to \cdots \to \tilde{H}^{0}(G, A) \to \cdots$$
$$\to \tilde{H}^{n}(G, A'') \to \tilde{H}^{n+1}(G, A') \to \cdots$$

extending from $-\infty$ *to* ∞ *is exact. Moreover, from a small commutative diagram*

$$\begin{array}{ccccccccc} 0 & \longrightarrow & A' & \longrightarrow & A & \longrightarrow & A'' & \longrightarrow & 0 \\ & & \downarrow & & \downarrow & & \downarrow & & \\ 0 & \longrightarrow & B' & \longrightarrow & B & \longrightarrow & B'' & \longrightarrow & 0 \end{array}$$

one obtains a doubly infinite commutative exact diagram

$$\begin{array}{ccccccccccc} \cdots \to & \tilde{H}^{-n}(G, A'') & \to & \tilde{H}^{-n+1}(G, A') & \to \cdots \to & \tilde{H}^{n}(G, A'') & \to & \tilde{H}^{n+1}(G, A') & \to \cdots \\ & \downarrow & & \downarrow & & \downarrow & & \downarrow & \\ \cdots \to & \tilde{H}^{-n}(G, B'') & \to & \tilde{H}^{-n+1}(G, B') & \to \cdots \to & \tilde{H}^{n}(G, B'') & \to & \tilde{H}^{n+1}(G, B') & \to \cdots \end{array}$$

There are other important facts: If $g = |G|$, then $g\tilde{H}^{n}(G, A) = (0)$ for all n between $-\infty$ and ∞; and if A is m-torsion, so are the groups $\tilde{H}^{n}(G, A)$ for $-\infty < n < \infty$. For our purposes, we need two more theorems.

THEOREM 17. (Duality Theorem). *Let* A *be a* G*-module for the finite group* G*. Then there is an isomorphism*

(34) $$\tilde{H}^{r-1}(G, A^{D}) \xrightarrow{\ \sim\ } \tilde{H}^{-r}(G, A)^{D}, \quad \textit{all } r \in \mathbf{Z}$$

where $(\ \ )^D$ *means the Pontrjagin dual of* $(\ \ )$.

THEOREM 18. (Integral Duality Theorem). *If* G *is a finite group, there is an isomorphism*

$$\tilde{H}^{-r}(G,Z) \xrightarrow{\ \sim\ } \tilde{H}^{r}(G,Z)^D, \quad \textit{all } r \in Z. \tag{35}$$

The important point in these theorems is that these dualities are given by cup-products; hence, they are functorial.

We can now resume our discussion of the Šaferevič-Golod Theorem. It depends upon two statements, of which the first is

LEMMA 4. (D. S. Rim). *Let* G *be a finite* p*-group, and let* $\Lambda = Z[G]$ *be its integral group ring. Let* A *be a* G*-module for which* $pA = (0)$, *let* $\Lambda/p\Lambda$ *be the group ring over* Z/pZ, *and let* I *be the augmentation ideal of* Λ (i.e., *the ideal generated by the elements* $\sigma - 1$ *as* σ *ranges over* G). *Then the minimal number of generators of* A *as* G*-module is the dimension of* $H_0(G, A) = A/IA$ *over* Z/pZ. *That is,* $a_1, \ldots, a_n$ *generate* A *as* G*-module if and only if* $\bar{a}_1, \ldots, \bar{a}_n$ *generate* A/IA *as vector space over* Z/pZ.

Proof: Clearly, if $a_1, \ldots, a_n$ generate A as G-module, then $\bar{a}_1, \ldots, \bar{a}_n$ generate A/IA over Z/pZ; it is the converse which requires proof. Let $\bar{a}_1, \ldots, \bar{a}_n$ generate A/IA as vector space over Z/pZ, and let B be the submodule of A generated by the pre-images $a_1, \ldots, a_n$ in A. We have the exact sequence

$$0 \to B \to A \to A/B \to 0 \ ;$$

hence, the homology sequence

$$H_0(G,B) \to H_0(G,A) \to H_0(G,A/B) \to 0 \ .$$

However, by hypothesis, the mapping $H_0(G,B) \to H_0(G,A)$ is surjective; hence, $H_0(G, A/B) = (0)$. Now A/B has a composition series all of whose factors are Z/pZ with trivial G-action, so were $A/B \neq (0)$, induction on the length of such a series would show that $H_0(G, A/B) \neq (0)$, a contradiction. It follows that $A/B = (0)$, that is, $A = B$. Q.E.D.

Lastly we need,

LEMMA 5. *Let* G *be a finite* p*-group, and let* A *be a* G*-module with* $pA = (0)$. *Then there exists a resolution (that is, an exact sequence)*

$$\cdots \to Y_3 \to Y_2 \to Y_1 \to Y_0 \to A \to 0$$

with the following properties

(a) *Each* Y_n *is free over* $\Lambda/p\Lambda$

(b) *The number of free generators of* Y_n *over* $\Lambda/p\Lambda$ *is precisely the dimension of* $H_n(G, A)$ *over* $\mathbb{Z}/p\mathbb{Z}$

(c) *The image of* Y_{n+1} *in* Y_n *is contained in* IY_n.

Proof: Let $d = \dim H_0(G, A)$, then by Rim's Lemma d is the minimal number of generators for A as G-module, and there exists a free Λ-module of rank d, say X, such that we have an exact sequence

$$X \to A \to 0 .$$

Let Y_0 be the cokernel of the mapping $X \xrightarrow{p} X$. Then Y_0 is a free $\Lambda/p\Lambda$-module of rank d.

Now, I claim $H_n(G, Y_0) = (0)$ for positive n. To see this, write

$$0 \to X \xrightarrow[p]{} X \to Y_0 \to 0$$

and apply homology:

$$\cdots \to H_{n+1}(X) \to H_{n+1}(Y_0) \to H_n(X) \to \cdots \to H_1(X) \to H_1(Y_0)$$

$$\to H_0(X) \xrightarrow[p]{} H_0(X) \to H_0(Y_0) \to 0 .$$

Since X is Λ-free, the groups $H_n(X)$, for $n > 0$, vanish; hence, $H_n(Y_0) = (0)$ for $n \geqq 2$. Moreover, the above exact sequence shows that

$$0 \to H_1(Y_0) \to X/IX \xrightarrow{p} X/IX$$

is exact. But X/IX is the direct sum of d copies of $\mathbb{Z}$; so it is torsion free. It follows that $H_1(Y_0) = (0)$, as claimed.

Let $B = \ker(Y_0 \to A)$, so that

$$0 \longrightarrow B \longrightarrow Y_0 \longrightarrow A \longrightarrow 0$$

is exact. Rim's Lemma and the construction of Y_0 show that the surjection $H_0(Y_0) \to H_0(A)$ is also an injection. Hence, the homology sequence becomes

$$\cdots \to 0 \to H_{n+1}(A) \to H_n(B) \to 0 \to \cdots \to 0 \to H_1(A) \to H_0(B) \to 0,$$

that is, $H_{n+1}(A) = H_n(B)$ for all $n \geq 0$. Since $H_0(Y_0) = H_0(A)$, the mapping

$$B/IB = H_0(B) \longrightarrow H_0(Y_0) = Y_0/IY_0$$

is the zero map, from which it follows that $B \subseteq IY_0$.

We now apply the above procedure to B instead of A (observe that $pY_0 = (0)$ implies $pB = (0)$), and obtain the free module Y_1 and the exact sequence

$$0 \to C \to Y_1 \to B \to 0 \ .$$

Moreover, $H_n(C) \cong H_{n+1}(B) \cong H_{n+2}(A)$, all $n \geq 0$; and $C \subseteq IY_1$. The module Y_1 has rank equal to the dimension of $H_0(B)$ which is the same as the dimension of $H_1(A)$, and the mapping $Y_1 \to Y_0$, obtained by composition $Y_1 \to B \to Y_0$, has image B which is contained in IY_0. From these remarks, we see that our procedure can be extended by induction, and it yields the desired resolution. Q.E.D.

Now we can prove Šafarevič-Golod's Theorem.

THEOREM 19. (Šafarevič-Golod).*. *Let* G *be a finite* p- *group, then*

(36) $$h_2(G) > \tfrac{1}{4} h_1(G)^2 \ ;$$

hence,

$$\lim_{h_1(G) \to \infty} \underline{\sigma}(G) = \infty \ .$$

* The theorem in this strengthened form is due to Roquette, Vinberg, Gaschutz who improved the original proof of Šafarevič and Golod.

Proof: According to the duality theorem for the cohomology of finite groups (Theorem 17) as applied to the module Z/pZ, we have

$$\tilde{H}^{r-1}(G, Z/pZ) \xrightarrow{\sim} \tilde{H}^{-r}(G, Z/pZ)^D .$$

(Observe that $(Z/pZ)^D = Z/pZ$.) In particular, the dimension of $\tilde{H}^{r-1}(G, Z/pZ)$ is the same as that of $\tilde{H}^{-r}(G, Z/pZ)$. Set $r = 2$, then

$$h_1(G) = \dim \tilde{H}^{-2}(G, Z/pZ) = \dim H_1(G, Z/pZ) ,$$

and when $r = 3$, we get

$$h_2(G) = \dim \tilde{H}^{-3}(G, Z/pZ) = \dim H_2(G, Z/pZ) .$$

Therefore, the theorem in question concerns the behavior of the homology of G with coefficients in Z/pZ.

We apply Lemma 5 to the G-module Z/pZ. Since $H_0(G, Z/pZ) = Z/pZ$, which has dimension one over Z/pZ, we can find Y_0 of rank one over $\Lambda/p\Lambda$ so that $Y_0 \to Z/pZ \to 0$ is exact. By part (c) of Lemma 5, the kernel of $(Y_0 \to Z/pZ)$ is contained in IY_0. However, as vector spaces over Z/pZ, the groups Y_0/IY_0 and Z/pZ are isomorphic; consequently, Rim's Lemma implies that Y_0/IY_0 and Z/pZ are isomorphic G-modules. Thus

$$0 \to IY_0 \to Y_0 \to Z/pZ \to 0$$

is exact. It follows from Lemma 5 that

$$Y_2 \to Y_1 \to IY_0 \to 0$$

is exact. Let $Y_2 = R$, $Y_1 = D$, $Y_0 = E$, then R, D, E are free $\Lambda/p\Lambda$-modules of ranks $h_2(G)$, $h_1(G)$, and 1 respectively. Moreover the sequence

$$(*) \qquad R \to D \to IE \to 0$$

is exact.

Let us now introduce the *Poincaré polynomial* of a finite G-module A with $pA = (0)$. For this we form the graded group

$$gr(A) = \coprod_{n=0}^{\infty} I^n A/I^{n+1} A ,$$

where $I^nA/I^{n+1}A$ is $I^n/I^{n+1} \otimes_{\Lambda/p\Lambda} A$. One knows that I/I^2 is isomorphic to $G/[G,G]$; hence, because p-groups are solvable, there is an integer N such that $I^mA/I^{m+1}A = (0)$ for all $m > N$. Now the groups $I^nA/I^{n+1}A$ are vector spaces over Z/pZ; we let $c_n(A)$ be the dimension of $I^nA/I^{n+1}A$ over Z/pZ. The Poincaré polynomial of A is defined by

$$\chi_A(t) = \sum_{n=0}^{\infty} c_n(A)t^n \quad . \tag{37}$$

Set $\chi(t)$ equal to $\chi_E(t)$, and observe trivially that if F is the free $\Lambda/p\Lambda$-module of rank n, then $\chi_F(t) = n\chi(t)$. From this, we deduce that

$$\chi_D(t) = h_1(G)\chi(t)\ ; \quad \chi_R(t) = h_2(G)\chi(t) \quad .$$

Moreover, $c_0(E) = \dim E/IE = 1$, and $c_n(IE) = \dim I^nIE/I^{n+1}IE$ $= \dim I^{n+1}E/I^{n+2}E = c_{n+1}(E)$; so one finds

$$\chi_{IE}(t) = \frac{\chi(t)-1}{t} \quad .$$

More generally, if $0 < t < 1$ is a real variable, then

$$\chi_A(t)\frac{1}{1-t} = \chi_A(t)\sum_{j=0}^{\infty} t^j = \sum_{m=0}^{\infty} c_m(A)t^m \sum_{j=0}^{\infty} t^j$$

$$= \sum_{r=0}^{\infty}\left(\sum_{m+j=r} c_m(A)\right)t^r = \sum_{r=0}^{\infty} s_r(A)t^r \quad .$$

Here, $s_r(A) = \Sigma_{m=0}^{r} c_m(A)$; hence, one sees that, $s_r(A) = \dim A/I^{r+1}A$ over Z/pZ. If we set $s_{-1}(A) = 0$, we can write

$$\chi_A(t)\frac{t}{1-t} = \sum_{r=0}^{\infty} s_{r-1}(A)t^r \; .$$

Since sequence (*) is exact, we have a *surjection* $I^{n+1}D \to I^{n+2}E$. If R_{n+1} is the inverse image of $I^{n+1}D$ in R, then the sequence

$$0 \to R/R_{n+1} \to D/I^{n+1}D \to IE/I^{n+2}E \to 0$$

is exact. Consequently,

$$s_n(D) = s_n(IE) + \dim_{Z/pZ}(R/R_{n+1}) \ .$$

Lemma 5 says that the image of R in D is contained in ID, so the image of I^nR is contained in $I^{n+1}D$. It follows that $I^nR \subseteq R_{n+1}$, so that $\dim R/R_{n+1} \leqq \dim(R/I^nR) = s_{n-1}(R)$. We deduce that

$$s_n(D) \leqq s_n(IE) + s_{n-1}(R) \ .$$

Going to our Poincaré power series for $0 < t < 1$, we may then write

$$\chi_D(t)\frac{1}{1-t} \leqq \chi_{IE}(t)\frac{1}{1-t} + \chi_R(t)\frac{t}{1-t} \ .$$

Hence,

$$h_1(G)\chi(t)\frac{1}{1-t} \leqq \frac{\chi(t)-1}{t}\cdot\frac{1}{1-t} + h_2(G)\chi(t)\frac{t}{1-t} \ .$$

From this, we deduce

$$1 \leqq \chi(t)\,(1-t\,h_1(G)+t^2\,h_2(G)), \quad 0 < t < 1 \ . \tag{38}$$

(Equation (38) is sometimes called Golod's Lemma.) But $\chi(t)$ is a positive function for positive values of t, and equation (38) yields the crucial inequality

$$1 - t\,h_1(G) + t^2 h_2(G) > 0 \quad \text{for } 0 < t < 1 \ .$$

As $h_2(G) \geqq h_1(G)$, we deduce $2h_2(G) > h_1(G)$, so that $0 < \dfrac{h_1(G)}{2h_2(G)} < 1$. Putting $t = \dfrac{h_1(G)}{2h_2(G)}$ in the above inequality, we obtain

$$h_2(G) > \frac{1}{4}\,h_1(G)^2,$$

as required. Q.E.D.

Remarks. The beautiful proof presented above has several interesting features. In the first place, the ring $\Lambda/p\Lambda$ is a non-commutative local

Artin ring and we were studying modules over it. Secondly, the Poincaré polynomial is nothing else than the non-commutative analog of the Hilbert-Samuel polynomial of algebraic geometry. Moreover, Rim's Lemma is just the non-commutative Nakayama's Lemma, and the whole question of generators and relations is very similar to the Hilbert-Zyzygy theory. From these remarks, it is very clear that the proper interpretation of the Šafarevič-Golod Theorem is as a theorem of "non-commutative algebraic geometry"—and we hope that the foundations of this subject will be quickly developed so that its rich applications as exemplified above will become available.

§4. Šafarevič's solution of the class tower problem

This section concerns global class field theory which will not be treated in the notes below. The central result is the negative solution to a long outstanding conjecture of Fürtwängler on class towers—provided by Šafarevič in 1962 [S], modulo the conjecture which he and Golod settled affirmatively in the Šafarevič-Golod Theorem (1964). This is such an outstanding application of Theorem 19, that we cannot resist giving it here, even though it requires some reasonable knowledge of class field theory to understand it. We shall provide very sketchy explanatory material to help orient the reader, but we advise readers having no knowledge of class field theory to skip this section and return to it later on.

We need two preparatory results.

PROPOSITION 28. *Let* G *be a profinite group (not necessarily a* p-*group). Then there exists in* G *a closed normal subgroup* T_p *such that* $G(p) = G/T_p$ *is a* p-*group, and such that if* G/N *is a* p-*group (where* N *is closed and normal in* G*), then* $T_p \subseteq N$. *That is,* G(p) *is the* maximal p-*quotient of* G.

Proof: As usual, let $\mathfrak{N}$ be the set of all closed, normal subgroups, N, of G for which G/N is a p-group. Since $G \in \mathfrak{N}$, $\mathfrak{N}$ is non-empty. If $\{N_\alpha\}$

is a chain in $\mathfrak{N}$, then $N = \cap N_\alpha$ is closed and normal. Hence, as $G/N = \operatorname{proj}_\alpha \lim G/N_\alpha$ so is a p-group, $\mathfrak{N}$ is inductive. Zorn's Lemma provides us with a minimal element of $\mathfrak{N}$, say T_p. If N is a closed normal subgroup such that G/N is a p-group, then

$$\begin{aligned}(G: T_p \cap N) &= (G:T_p)(T_p:T_p \cap N)\\ &= (G:T_p)(T_pN:N)\ .\end{aligned}$$

But $(G:T_p)$ is a p-power, and $(T_pN:N)$ divides $(G:N)$ which is a p-power. Thus $G/T_p \cap N$ is a p-group, so by the minimality of T_p, we deduce $T_p \cap N \supseteqq T_p$. Thus, $N \supseteqq T_p$. Q.E.D.

By a *number field* we mean a finite extension of the field of rational numbers. Global class field theory is concerned with *abelian* extensions of number fields. It associates to each number field k, a topological group, C_k, the group of *idèle classes* of k. Then it classifies the abelian extensions K over k by associating with each such K/k, the group $C_k/\mathfrak{N}_{K/k}C_K$, where $\mathfrak{N}$ is the norm mapping of field theory. The association is all the more remarkable because of *Artin's reciprocity law* which says that a certain canonically defined map $G(K/k) \to C_k/\mathfrak{N}_{K/k}\, C_K$ is an *isomorphism*—where $G(K/k)$ is the Galois group of K/k.

Slightly more explicitly, we let I_k—the idèle group of k—be defined by

$$I_k = \{\xi \in \prod_P k_P^* \mid \text{for almost all } P,\ \xi_P (= P\text{-component of } \xi)$$
$$\text{is a unit of } k_P^*,\ \text{i.e., } |\xi_P|_P = 1 \text{ for almost all } P\}\ .$$

Here, P ranges over all the *primes* of k.
Then C_k is I_k/k^*, and we let U_k be the subgroup of I_k consisting of those $\xi \in I_k$ for which ξ_P is a unit whenever P is a non-archimedian prime of k. The intersection $k^* \cap U_k$, denoted E_k, is called the *group of units of* k. The group E_k contains the group of all roots of unity contained in k (a finite group), and the famous Dirichlet-Minkowski-Hasse-Chavalley Unit Theorem asserts that *the factor group of* E_k *by the roots of unity is a free abelian group of rank* $s-1$, *where* s *is the number of archimedian primes*

of k. It follows that E_k, as abelian group, is generated by s elements. One also has the *ideal class* group of k, denoted Cl_k, which is defined by

$$Cl_k = I_k/k^* U_k = C_k/(U_k/E_k) \ .$$

In the classification of abelian extensions, described above, corresponding to factor groups of C_k, *those which are actually factor groups of* Cl_k *are precisely the unramified extensions of* k *(abelian, of course!)* It follows from this that a *number field* k *possesses a maximal, abelian, unramified extension,* called its *Hilbert Class Field.*

We have the exact sequences

$$0 \longrightarrow U_k/E_k \longrightarrow C_k \longrightarrow Cl_k \longrightarrow 0$$

$$0 \longrightarrow E_k \longrightarrow U_k \longrightarrow U_k/E_k \longrightarrow 0 \ ,$$

and a study of the cohomology of these sequences gives the connection between class field theory and Šafarevič's number.

PROPOSITION 29. (Iwasawa [I]). *Let* k *be a number field, and let* K/k *be an unramified,* p*-extension (i.e.,* K/k *is a Galois extension and* G(K/k) *is a* p*-group). Suppose that* K *has no cyclic, unramified extension of degree* p. *If* r_1 *(resp.* r_2*) denotes the number of real (resp. complex) conjugates of* k, *then*

$$\underline{\sigma}(G) \leqq r_1 + r_2 \ , \quad \textit{where } G = G(K/k) \ .$$

Proof: We will use the exact sequences of G-modules

$$0 \longrightarrow U_K/E_K \longrightarrow C_K \longrightarrow Cl_K \longrightarrow 0$$

$$0 \longrightarrow E_K \longrightarrow U_K \longrightarrow U_K/E_K \longrightarrow 0 \ .$$

The assumption that K possesses no cyclic unramified p-extensions is precisely the statement that Cl_K (*a finite group by the finiteness of class number*) has order prime-to-p. This follows because Cl_K classifies the abelian unramified extensions of K, and were p to divide $|Cl_K|$, it would follow that Cl_K has a factor group of order p —hence, that K has an unramified, cyclic extension of degree p. Since G is a p-group and since Cl_K has order prime-to-p, it follows, from the discussion just before Theorem 17, that

$$\tilde{H}^n(G, Cl_K) = (0) \quad , \qquad \text{for all } n \in Z .$$

Moreover, it is known from class field theory, that as K/k is unramified, the cohomology groups $\tilde{H}^n(G, U_K)$ vanish for every n. (We will prove this in Chapter V.) Cohomology applied to our two sequences therefore yields

$$\tilde{H}^r(G, U_K/E_K) \xrightarrow{\sim} \tilde{H}^r(G, C_K), \qquad \text{for all } r \in Z$$
$$\tilde{H}^r(G, U_K/E_K) \xrightarrow{\sim} \tilde{H}^{r+1}(G, E_K), \quad \text{for all } r \in Z .$$

Thus, $\tilde{H}^r(G, C_K) \cong \tilde{H}^{r+1}(G, E_K)$ for every $r \in Z$. Now the main theorem of class field theory (in Tate's form) states that a certain canonical mapping $\tilde{H}^{r-2}(G, Z) \to \tilde{H}^r(G, C_K)$ is an *isomorphism* for all $r \in Z$. If we put this together with the above, we deduce

$$\tilde{H}^{r-2}(G, Z) \cong \tilde{H}^{r+1}(G, E_K) .$$

Let $r = -1$, then $\tilde{H}^{-3}(G, Z) \cong \tilde{H}^0(G, E_K) = E_k/\mathfrak{N}_{K/k} E_K$. But the integral duality theorem (Theorem 18) shows that this implies

$$\text{rank}\,(\tilde{H}^3(G, Z)) = \text{rank}\,(E_k/\mathfrak{N}_{K/k} E_K) .$$

From Proposition 27 and the fact that $r_1 + r_2$ is the number of archimedian primes in k (so that by the Dirichlet Unit Theorem E_k has $r_1 + r_2$ generators), we finally deduce

$$\underline{\sigma}(G) \leqq r_1 + r_2 . \qquad\qquad \text{Q.E.D.}$$

Now to the class tower problem. We have already mentioned that each number field possesses a finite, maximal, abelian, unramified extension, called its Hilbert Class Field (HCF). The degree of HCF (k) over k is exactly the order of Cl_k, and is called the *class number* of k. If a number field has class number one, then its ring of integers (always a Dedekind domain) is a PID, and conversely. That is, class number one is synonomous with unique factorization in the ring of integers of k. Moreover, there is a famous theorem—the *principal ideal theorem*—which states that every prime ideal of k becomes principal in HCF (k). From these remarks, it is evident that the field HCF (k) is an important invariant for the arithmetic structure of k. Fürtwängler, in 1924, considered the following problem—the so-called *class tower problem: Let* k *be a number field, and let* $k_1 = \text{HCF}(k)$. *Form the field* $k_2 = \text{HCF}(k_1)$, *and continue in this fashion. One obtains a tower of fields*

$$k \subseteq k_1 \subseteq k_2 \subseteq k_3 \subseteq \cdots$$

each the Hilbert Class Field of the field preceding it. One asks: Does this tower of Hilbert Class Fields stop? Observe that each k_j is unramified over k (but NOT abelian, in general), and that the union of the k_j is the maximal unramified extension of k.

THEOREM 20 (Šafarevič). *The class tower problem admits a negative answer. That is, there do exist algebraic number fields whose class towers are infinite. For example, the field*

$$k = \mathbf{Q}(\sqrt{-3 \cdot 5 \cdot 7 \cdot 11 \cdot 13 \cdot 17 \cdot 19})$$

has an infinite class tower above it. Consequently, the class tower problem even admits a negative answer for an imaginary quadratic field.

Proof: Let k be a number field, let Ω be its maximal unramified extension, and let $G = G(\Omega/k)$ be the Galois group. Then (as we shall show, in Chapter IV, §1), G has a natural topology (the Krull topology) under which it becomes a profinite group. Let G(p) be its maximal p-quotient for some prime number p. The group G(p) corresponds, by Galois theory, to the maximal unramified p-extension of k, say k(p). If we show that there exists a field k for which k(p)/k is an infinite extension, it will follow from the remarks above that Ω/k is infinite and that the class tower of k does not stop.

Therefore, we may and do assume that $\Omega = k(p)$, and $G = G(p)$. If

$p = 2$, we can assume k is totally imaginary. Now suppose the theorem is false. Then for every algebraic number field, k, the extension $k(p)/k$ is a *finite* extension. Consequently, G is a finite p-group; and as $k(p)$ admits no cyclic unramified extensions of degree p, Proposition 29, shows that

$$\underline{\sigma}(G) \leqq r_1 + r_2 \leqq [k:\mathbf{Q}] < \infty .$$

Now take $p = 2$, k totally imaginary, and consider $p_1, \ldots, p_N$, distinct odd primes. Arrange N so that

$$-p_1 p_2 \cdots p_N \equiv 1 \bmod 4 ,$$

e.g., -3, -3.5, $3 \cdot 5 \cdot 7$, etc. will do. Form the quadratic extension $k = \mathbf{Q}(\sqrt{-p_1 p_2 \cdots p_N})$, and let $K_j = k(\sqrt{\pm p_j})$, where we choose $+p_j$ if $p_j \equiv 1 \bmod 4$, and $-p_j$ if $p_j \equiv 3 \bmod 4$.

One sees evidently that the extensions $K_1, K_2, \ldots, K_N$ are independent, unramified extensions of k. Since $h_1(G) = h_1(G^{ab})$, and since G^{ab} is (by Artin's reciprocity law) isomorphic to $C_k/\mathfrak{N}_{k(2)/k}\, C_{k(2)}$, we deduce that

$$h_1(G) = h_1(C_k/\mathfrak{N}_{k(2)/k}\, C_{k(2)}) .$$

But G^{ab} is the Galois group of an *unramified* extension—in fact the maximal abelian unramified 2-extension of k. Consequently, our remarks preceding Proposition 29 show that

$$h_1(G) = h_1(G^{ab}) = h_1(Cl_k(2))$$

where $Cl_k(2)$ is the 2-primary part of Cl_k. However, we have already exhibited N *independent*, (abelian), unramified, quadratic extensions of k; hence, $Cl_k(2)$ has at least N generators. Thus, $h_1(G) \geqq N$. Because the field k is totally imaginary, $r_1 = 0$, and because $[k:\mathbf{Q}] = 2$, $r_2 = 1$. Thus, Proposition 29 shows that $\underline{\sigma}(G) \leqq 1$. However, N is arbitrary, so a contradiction of the Šafarevič-Golod Theorem has been reached. We conclude that the theorem holds.

To justify the explicit counter-example, note that $h_2(G) > h_1(G)^2/4$;

so if we want to contradict $h_2(G) - h_1(G) \leqq 1$, we need only take $N \geqq 6$ to do it. In our example, we have chosen $N = 7$. Q.E.D.

COROLLARY. *The embedding problem also has a negative answer. That is, there exist algebraic number fields,* k, *with class number greater than one, which cannot be embedded in finite algebraic number fields whose class number is one.*

We leave the proof of this as an exercise.

CHAPTER IV

GALOIS COHOMOLOGY AND FIELD THEORY

§1. Resume of Krull's Galois Theory and First Results

Let k be a field and let k_s be its separable closure. Then k_s is a normal extension of k, and we may write $k_s = \text{dir lim } K$ where K runs over the *finite, normal, separable* extensions of k. Let $G_k = G(k_s/k)$ denote the group of all k-automorphisms of k_s, that is, the Galois group of k_s/k. Let $\{K_\alpha\}$ denote the family of finite, normal, separable extensions of k, and let U_α be the invariance group of K_α in G_k; that is,

$$U_\alpha = \{\sigma \in G_k \mid \sigma \mid K_\alpha = 1\} .$$

Ordinary Galois-theoretic arguments show that each U_α is normal in G_k and that $G(K_\alpha/k) = G_k/U_\alpha$; so that each U_α has finite index in G_k. Let $\mathfrak{B}$ denote the set of all these U_α and take them as a fundamental system of neighborhoods of 1 in G_k. This gives G_k a topology, called the *Krull Topology.*

THEOREM 21 (Krull). *Let* G_k *be the Galois group of* k_s/k *with the Krull topology. Then* G_k *is the projective limit of the groups* $G(K_\alpha/k)$; *hence,* G_k *is a profinite group. The usual correspondence of Galois theory is a* 1-1 *lattice inverting correspondence between the set of all closed subgroups of* G_k *and the set of all subextensions of* k_s/k. *Finite extensions correspond to open subgroups, and the usual subgroup and factor group correspondences hold. The same statements hold if* k_s *is replaced by any normal, separable, algebraic extension of* k *and* G_k *is replaced mutatis-mutandis*

We shall leave the proof to the reader as an exercise.

If K/k is a Galois extension (not necessarily finite), we wish to study the cohomology groups of $G(K/k)$. To emphasize the point of view that these groups are invariants of the field extension K/k, we shall adopt different notation for the cohomology as follows:

For a module A in the appropriate category, we set

$$H^r(K/k, A) = H^r(G(K/k), A)$$

$$H^r(k, A) = H^r(G(k_s/k), A) = H^r(G_k, A).$$

In this notation, we let $A(K) = A^{G_K}$, and we may then write the Hochschild-Serre Spectral Sequence in the form

$$H^p(K/k, H^q(K, A)) \Longrightarrow H^*(k, A) \ .$$

The exact sequence of terms of low degree becomes

$$0 \to H^1(K/k, A(K)) \to H^1(k, A) \to H^1(K, A)^{G(K/k)} \to H^2(K/k, A(K)) \to H^2(k, A) \ .$$

We shall say that the field k has $(s)\mathrm{cd}_p \leqq n$ (or $(s)\,\mathrm{cd}_p = n$), and write $(s)\,\mathrm{cd}_p\, k \leqq n$ $(= n)$, if and only if the same is true of the Galois group G_k. For the extension K/k, there are two extremely important K/k-modules (i.e., $G(K/k)$-modules):

(a) K^+ – the additive group of K;

(b) K^* – the multiplicative group of K.

Both are acted on continuously by $G(K/k)$ in the natural action, and the basic result is

THEOREM 22 (Hilbert Theorem 90). *Let* K/k *be a Galois extension, then*

$$H^r(K/k, K^+) = (0) \qquad \text{for } r > 0$$

and

$$H^1(K/k, K^*) = (0) \ .$$

Proof: $H^r(K/k, K^+)$ is the direct limit under inflations of the groups $H^r(L/k, L^+)$ for finite, normal extensions L/k. Therefore, we may assume K/k is a finite extension. The normal basis theorem asserts that there is an element $\theta \in K$, such that the set $\{\sigma\theta \mid \sigma \in G(K/k)\}$ is a k-basis for the vector space K^+. Consider the direct image $\pi_{*k\to K}k^+$ $(= \pi_{*\{1\}\to G(K/k)}k^+)$ of the module k^+ for the trivial group. If f is an element of this direct image, let $T(f) = \Sigma_{\sigma\in G(K/k)} f(\sigma)\sigma\theta$. Since the $\sigma\theta$ are linearly independent over k, the mapping T is a monomorphism. But $\dim_k(\pi_* k^+) = [K:k]$; hence, T is actually an isomorphism. Moreover, a simple computation shows that T is a $G(K/k)$-isomorphism. We deduce that

$$H^r(K/k, K^+) = H^r(K/k, \pi_{*k\to K}k^+) = H^r(k/k, k^+) = (0), \quad r > 0.$$

For the multiplicative part, we may again assume K/k is a finite extension. Let $f(\sigma)$ be a 1-cocycle representing a given element $a \in H^1(K/k, K^*)$. If $\xi \in K^*$, define $A(\xi)$ by

$$A(\xi) = \sum_{\sigma\in G(K/k)} f(\sigma)\sigma\xi\ ,$$

and note that as all the automorphisms of $G(K/k)$ are distinct, they are independent [A]; hence, there is a ξ for which $A(\xi) \neq 0$. Pick such a ξ, and set $b = A(\xi)^{-1}$. Then

$$\sigma b^{-1} = \sigma\Sigma_{\tau\in G(K/k)} f(\tau)\tau\xi = \Sigma_\tau\, \sigma f(\tau)\sigma\tau\xi = \Sigma_\tau \frac{f(\sigma\tau)}{f(\sigma)}(\sigma\tau)\xi$$
$$= \frac{1}{f(\sigma)}\, b^{-1}\ .$$

Hence, $f(\sigma) = \sigma b/b$; that is, f is a 1-coboundary. Q.E.D.

The group $H^2(k, k_s^*)$ is especially important, it is called the *Brauer Group of* k, denoted $Br(k)$, after Richard Brauer who first studied it systematically in a different guise [Br]. This group is a subtle arithmetic invariant of the field k. The group $H^2(K/k, K^*)$ will be denoted $Br(K/k)$, and will be called the *relative Brauer group* of K/k, or the *part of the Brauer group split by* K. There are important corollaries of Theorem 22 dealing with the Brauer group.

Corollary 1. *If* $L \supseteq K \supseteq k$ *are field extensions, then the sequence*

$$0 \longrightarrow Br(K/k) \xrightarrow{\text{inf}} Br(L/k) \xrightarrow{\text{res}} Br(L/K) \xrightarrow{K/k} H^3(K/k, K^*) \longrightarrow H^3(L/k, L^*)$$

is exact. In particular, the inflation maps $Br(K/k) \to Br(k)$ *are always injective; so the Brauer group of* k *is the union of all the relative Brauer groups of extensions* K/k *for which* K/k *is finite and Galois.*

Proof: The sequence in question is the exact sequence of terms of low degree in the Hochschild-Serre sequence

$$H^p(K/k, H^q(L/K, L^*)) \Longrightarrow H^*(L/k, L^*) \quad ,$$

because $E_2^{p,q} = (0)$ if $q = 1$ by Hilbert Theorem 90. Q.E.D.

COROLLARY 2. *Let* k *be a field of characteristic* $p > 0$. *Then* $cd_p k \leqq 1$, *and* $Br(k)$ *is divisible by* p.

Proof: Consider the Artin-Schreier mapping $\wp$ from k_s^+ to itself given by $\wp(x) = x^p - x$. This mapping is a G_k-homomorphism and is surjective because k_s is separably closed. The kernel of $\wp$ is the set of all $x \in k_s^+$ for which $x^p = x$, and this is well-known to be $\mathbf{Z}/p\mathbf{Z}$. So we see that

$$0 \longrightarrow \mathbf{Z}/p\mathbf{Z} \longrightarrow k_s^+ \xrightarrow{\wp} k_s^+ \longrightarrow 0$$

is exact. The cohomology sequence yields

$$\cdots \to H^r(k, k_s^+) \to H^{r+1}(k, \mathbf{Z}/p\mathbf{Z}) \to H^{r+1}(k, k_s^+) \to \cdots \quad ,$$

and it follows that $H^s(k, \mathbf{Z}/p\mathbf{Z}) = (0)$, for all $s \geqq 2$. Now let G_p be a p-Sylow group of G_k, and let k_p be its fixed field. By what we have just shown, $H^s(k_p, \mathbf{Z}/p\mathbf{Z}) = (0)$ for all $s \geqq 2$. Thus, $cd_p G_p = cd_p k_p \leqq 1$, and we know $cd_p k = cd_p k_p$.

For the statement concerning $Br(k)$, look at the sequence

$$0 \longrightarrow k_s^* \xrightarrow{\ p\ } k_s^* \longrightarrow k_s^*/k_s^{*p} \longrightarrow 0 \ .$$

The right hand extreme group is a p-torsion group; so by the first part of the proof and the cohomology sequence, we get

$$\mathrm{Br}(k) \xrightarrow{\ p\ } \mathrm{Br}(k) \longrightarrow H^2(k, k_s^*/k_s^{*p}) = (0) \ .$$

Q.E.D.

Remark. The exact sequence $0 \to Z/pZ \to k_s^+ \xrightarrow{\ \wp\ } k_s^+ \to 0$ shows that $H^1(k, Z/pZ) = k^+/\wp(k^+)$, if $\mathrm{ch}(k) = p > 0$.

§2. The cohomological dimension of fields

In the proof of Corollary 2 above we introduced the field k_p, the fixed field of G_p—a p-Sylow subgroup of G_k. Thus, k_s/k_p is a p-extension, and k_p/k is prime-to-p. If T_p is the normal subgroup of G_k whose quotient, G(p), is the maximal p-quotient of G_k, (cf. Proposition 28), we let k(p) be the fixed field of T_p. Then k(p) is a normal, separable extension of k, its *maximal* p-*extension*. We already know that $\mathrm{cd}_p k = \mathrm{cd}_p k_p \leqq 1$ if p is the characteristic of k; and we now wish to investigate $\mathrm{cd}_p G(p)$ and the case $p \neq$ characteristic of k.

LEMMA 6. *Let* G *be a profinite group, and assume* $\mathrm{cd}_p T_p \leqq 1$. *Then the inflation maps*

$$H^r(G(p), Z/pZ) \to H^r(G, Z/pZ)$$

are isomorphisms; hence, $\mathrm{cd}_p G(p) \leqq \mathrm{cd}_p G$.

Proof: We claim $H^1(T_p, Z/pZ)$ is always zero without any hypotheses on $\mathrm{cd}_p T_p$. For let $\xi \in H^1(T_p, Z/pZ)$, that is $\xi \in \mathrm{Hom}(T_p, Z/pZ)$. Let $V = \ker \xi \subseteq T_p$, and assume $V \neq T_p$ (that is, $\xi \neq 0$). As G acts continuously by conjugation on ξ, there are only finitely many conjugates of ξ by G. Let N be the intersection of the kernels of all these conjugates of ξ, then N is a closed normal subgroup of G; $N \subseteq V < T_p$. But it is clear that T_p/N is a p-group, so we see that G/N is a p-group, and this contradicts the minimality of T_p.

Now our hypothesis shows that $H^r(T_p, Z/pZ)$ vanishes for $r > 1$; so we have proved that $H^r(T_p, Z/pZ)$ vanishes for $r > 0$. Thus, the Hochschild-Serre sequence

$$H^r(G(p), H^s(T_p, Z/pZ)) \underset{r}{\Longrightarrow} H^*(G, Z/pZ) \ .$$

degenerates, and the conclusions follow. Q.E.D.

Question: Is it true in general that $cd_p\, G(p) \leqq cd_p\, G$?

PROPOSITION 30 (Kawada). *Let* k *be a field of characteristic* $p > 0$, *and let* k(p) *be its maximal* p-*extension. If* $G_k(p)$ *is the Galois group of* k(p) *over* k, *then* $G_k(p)$ *is a free profinite* p-*group on* $|k^+/\wp(k)^+|$ *generators.*

Proof: Since $T_p \subseteq G_k$, we see that $cd_p T_p \leqq cd_p k \leqq 1$. Lemma 6 yields: $cd_p\, G_k(p) \leqq 1$, so that $G_k(p)$ is indeed a free profinite p-group. Now the sequence

$$0 \longrightarrow Z/pZ \longrightarrow k(p)^+ \xrightarrow{\ \wp\ } k(p)^+ \longrightarrow 0$$

is exact; so we deduce

$$H^1(G_k(p), Z/pZ) \cong k^+/\wp(k^+) \ .$$

Our proposition follows from Corollary 2 of Proposition 23. Q.E.D.

PROPOSITION 31. (Albert-Hochschild) *Let* k *be a field of characteristic* $p > 0$, *and let* k′ *be a purely inseparable extension of* k. *Then the canonical "restriction" mapping* $Br(k) \to Br(k')$ *is surjective.*

Proof: Since k′/k is purely inseparable, the theorem on natural irrationalities [A] shows that $G_{k'} \to G_k$ is an *isomorphism*. We then consider the exact sequence

$$0 \longrightarrow k_s{}^* \xrightarrow{\ i\ } k_s'{}^* \longrightarrow \text{Coker } i \longrightarrow 0$$

of G_k-modules. If $\xi \in k_s'{}^*$, there exists a power p^r of the characteristic such that $\xi^{p^r} \in k_s{}^*$. This shows that coker i is a p-torsion module; hence

$$\text{Br}(k) \longrightarrow \text{Br}(k') \longrightarrow H^2(k, \text{Coker } i) = (0)$$

is exact. Q.E.D.

Now we shall investigate the case $p \neq \text{ch } k$.

THEOREM 23. *Let* k *be a field,* $\text{ch } k \neq p$, *then the following statements are equivalent:*

(a) $cd_p k \leqq n$ (for $n \geqq 1$)

(b) $H^{n+1}(K, K_s^*; p) = (0)$ *and* $H^n(K, K_s^*; p)$ *is p-divisible for all extensions* K/k *which are separable algebraic.*

(c) *Same as (2), except* K/k *is finite, separable and of degree prime-to-* p.

Proof: Let μ_p denote the group of pth roots of unity in k_s^*. We have the exact sequence

$$0 \longrightarrow \mu_p \longrightarrow k_s^* \xrightarrow{\ p\ } k_s^* \longrightarrow 0\ ;$$

hence, the cohomology sequence

$$\longrightarrow H^n(K, k_s^*; p) \xrightarrow{\ p\ } H^n(K, k_s^*; p) \longrightarrow H^{n+1}(K, \mu_p)$$
$$\longrightarrow H^{n+1}(K, k_s^*; p) \xrightarrow{\ p\ } H^{n+1}(K, k_s^*; p) \longrightarrow \cdots$$

Since $K_s^* = k_s^*$ for the types of extensions mentioned in (b) and (c), it follows trivially that (b) and (c) are equivalent to the statements: $H^{n+1}(K, \mu_p) = (0)$, where for (b) we choose all K/k as described in (b) and similarly for (c). With this remark in our possession we can now begin the proof.

(a) $\Longrightarrow$ (b). $cd_p k \leqq n$ implies $cd_p K \leqq n$ (as $G_K \subseteq G_k$); hence $H^{n+1}(K, \mu_p) = (0)$, and this proves (b).

(b) $\Longrightarrow$ (c). This is trivial.

(c) $\Longrightarrow$ (a). Let k_p be a p-Sylow field for k, then k_p is the direct limit: dir lim K, where the field extensions K/k satisfy the provisos of statement (c). It follows that

$$H^{n+1}(k_p, \mu_p) = \text{dir lim } H^{n+1}(K, \mu_p) = (0) \ ,$$

where the limit on the right is taken with respect to the restriction maps. Now $\mu_p \xrightarrow{\sim} Z/pZ$ as abelian groups, and in fact this is an isomorphism as $G_p = G(k_s/k_p)$-modules. (Because μ_p is a p-group, simple as G_p-module, and G_p is a p-profinite group.) [Our catch phrase for this will be: *μ_p is isomorphic to Z/pZ over k_p*.] Thus, $H^{n+1}(k_p, Z/pZ) = (0)$; hence, $cd_p k_p \leqq n$. But $cd_p k = cd_p k_p$, so the result follows. Q.E.D.

COROLLARY 1. *The pth roots of unity lie in every p-Sylow field k_p.*

COROLLARY 2. *If $p \neq$ ch k, then $cd_p k \leqq 1$ if and only if $Br(K; p) = (0)$ for all K/k finite, separable, and of degree prime-to-p.*

Proof: Combine Hilbert's Theorem 90 with the case $n = 1$ of the theorem above.

COROLLARY 3. *If k is a perfect field of characteristic $p > 0$, then $Br(k; p) = (0)$. Hence, for perfect fields k, we have*

$$Br(K; q) = (0) \Longleftrightarrow cd_q k \leqq 1$$

where K ranges over all finite, separable extensions of k. Moreover, we may write (if k is perfect)

$$cd\, k \leqq 1 \Longleftrightarrow Br(K) = (0) \text{ for all } K/k \text{ finite and separable.}$$

Proof: The sequence $0 \longrightarrow k_s^* \xrightarrow{p} k_s^* \longrightarrow 0 \longrightarrow 0$ is exact; so the mapping

$$Br(k) \xrightarrow{p} Br(k)$$

is an isomorphism. It follows that $Br(k; p) = (0)$. The rest of the argument now follows from Corollary 2. Q.E.D.

COROLLARY 4. *If* k *is a finite field or an absolutely algebraic field of positive characteristic, then* Br(k) = (0).

Proof: For the k's of the hypothesis, G_k is either $\hat{Z} = \text{proj lim } Z/n!Z = \prod_p Z_p$, or a closed subgroup of this group. (Here, $Z_p = \text{proj lim } Z/p^rZ$.) However, the p-Sylow group of $\hat{Z}$ is Z_p, and one sees that Z_p is the free profinite p-group on one generator. Thus $cd_p\, k \leqq 1$, for every p and Corollary 3 completes the proof. Q.E.D.

§3. Algebraic dimension, Quasi-algebraic Closure, and the property C_r

We shall introduce a notion for fields which is slightly stronger than cd k $\leqq$ 1, but is equivalent to this for perfect fields. For later work it will be convenient to have alternative characterizations of this new notion. One of them will involve *cohomological triviality*. We say a module A for the finite group G is *cohomologically trivial* if and only if $\tilde{H}^r(G, A) = (0)$, for every $r \in Z$. There is a criterion for cohomological triviality—sometimes called the "twin number" theorem (or like phraseology) – it goes as follows: *The module* A *is cohomologically trivial if and only if for every prime* p, *we can find consecutive numbers* i_p, $i_p + 1$ *such that* $\tilde{H}^{i_p}(G_p, A) = \tilde{H}^{i_p+1}(G_p, A) = (0)$, *where* G_p *is a* p-*Sylow subgroup of* G. A proof of this theorem will be given in Chapter V, §2.

PROPOSITION 32. *Let* k *be a field, then the following conditions are equivalent:*

(a) cd k $\leqq$ 1 *and if* ch k = p > 0, *also assume* Br(K ; p) *vanishes for all algebraic, separable extensions* K/k.

(b) Br(K) = (0) *for all separable, algebraic extensions* K/k.

(c) *If* K/k *is separable algebraic, and* L/K *is a finite Galois extension, then* L* *is a cohomologically trivial* G(L/K)-*module.*

(d) *If* K/k, L/K *are as in* (c), *then the norm mapping* $\mathfrak{N}_{L/K}: L^* \to K^*$ *is surjective.*

Proof: (a) $\Longrightarrow$ (b). This is essentially Corollary 2 to Theorem 23.

(b) $\Longrightarrow$ (c). Pick a Sylow subgroup, say G_p, for $G(L/K)$ and let K_p be its fixed field. Then K_p/k is separable, algebraic, and $H^2(L/K_p, L^*) \subseteqq Br(K_p) = (0)$, while $H^1(L/K_p, L^*) = (0)$ by Hilbert Theorem 90. The twin number criterion (for $i_p = 1$, $i_p + 1 = 2$) yields (c).

(c) $\Longrightarrow$ (d). By cohomological triviality, $\tilde{H}^0(L/K, L^*) = (0)$. Thus $K^*/\mathfrak{N}_{L/K} L^* = (0)$, that is, (d) holds.

(d) $\Longrightarrow$ (a). Assumption (d) is merely $\tilde{H}^0(L/K, L^*) = (0)$ for every pair L, K as described. Hilbert Theorem 90 shows that $H^1(L/K, L^*) = (0)$ for all such pairs. By the twin number criterion (for $0, 1$) we deduce that L^* is cohomologically trivial; so in particular, $H^2(L/K, L^*) = \tilde{H}^2(L/K, L^*) = (0)$. Since $Br(K) = \operatorname{dir\,lim}_{L} H^2(L/K, L^*)$, we see that (b) holds. But (b) implies (a) by Corollary 2 of Theorem 23. Q.E.D.

We will say that a field k has (*algebraic*) *dimension* $\leqq 1$ if and only if one of the above (hence all of the above) conditions holds for k. We will write $\dim k \leqq 1$ for this concept.

Observe that if K/k is separable algebraic, and if $\dim k \leqq 1$, then $\dim K \leqq 1$ as well. Also, Corollary 2 of Theorem 23 says that: *for perfect fields,* $\mathrm{cd}\, k \leqq 1 \Longleftrightarrow \dim k \leqq 1$. *If* k *is a field of dimension less than or equal to one, then for any prime* p, *the Galois group of* $k(p)/k$ *is a free profinite* p-*group.* To see this, observe that $\dim k \leqq 1$ implies $\dim k(p) \leqq 1$. This in turn implies $\mathrm{cd}_p\, k(p) \leqq 1$, and now Lemma 6 shows that $\mathrm{cd}_p\, G(p) \leqq \mathrm{cd}_p\, k \leqq 1$, as contended.

In the hierarchy of strength, the requirement "$\dim k \leqq 1$" is majorized by one of a series of requirements on k. The latter requirements have direct arithmetic and geometric meaning, and were introduced by E. Artin and S. Lang. Artin was inspired by a theorem of Tsen (see below). These conditions have fascinated many mathematicians, especially since Artin formulated several tempting conjectures in their language. Some of these conjectures were proven soon after being proposed, others were shown false as stated but true after modification, and still others await proofs or counterexamples. Here is the definition.

Definition 6. A field k *has property* C_r if and only if every form $f(x_1,\ldots,x_n)$ of degree d with coefficients in k has a non-trivial zero in k whenever $n > d^r$. A field is *quasi-algebraically closed* (QAC) iff it has C_1.

Remarks (1) Property C_r should be considered as a measure of how far k is from being algebraically closed. Indeed, k *has* C_0 *if and only if* k *is algebraically closed;*

(2) If k has C_r then k has C_s for every $s \geqq r$.

(3) Part of the fascination of property C_r is that it is a direct link between the algebraic geometry over the field k and the arithmetic structure of k. Thus, C_r is a property of "Diophantine geometry" in the generalized sense of Lang [LD, LQ].

(4) To indicate the interest of mathematicians in such questions, we shall state but not prove a theorem of Birch and Peck concerning forms over a number field. This is a qualitative, rather than a quantitative result, in contradistinction to the sort of theorems we will obtain below concerning the property C_r. The interested reader should consult [Bi, Gg, Pe] for more details.

Theorem (Birch, Peck). *If* k *is a number field and* f *is a form of degree* d *in* n *variables, then if either* d *is odd or* k *is totally imaginary,* f *has a zero in* k *provided* n *is sufficiently large.*

If K/k is a finite, normal, field extension of degree n, introduce n indeterminates $X_1,\ldots,X_n$ and choose a basis for K/k, say $\omega_1,\ldots,\omega_n$. The element $\Sigma X_j\omega_j$ may be thought of as the "generic element" of K, and we may form the norm of $\Sigma X_j\omega_j$:

$$\mathfrak{N}_{K/k}(\Sigma X_j\omega_j) = \prod_{\sigma\in G(K/k)} \Sigma X_j\omega_j^\sigma = \mathfrak{N}_{K/k}(X_1,\ldots,X_n)\,.$$

This is a form of degree n in n-variables called the *norm form.* Observe that the norm form has only the trivial zero in k. Based on this experience, let us call a form $f(X_1,\ldots,X_n)$ of degree d a *normic form of order* i iff $n = d^i$ and f has only the trivial zero in k. The form $f(X) = aX$ is a normic form of order i (for each i) and is of degree 1. We'll call it the trivial normic form.

Now there is a simple, but fruitful way of making new forms from old. Let f, g be forms, and introduce the notation $f(g|g|g|\cdots|g)$ to mean wherever a variable of f occurs substitute g and use new variables in g after each occurrence of $|$. If f has degree d and is a form in n variables, and if g has degree e and is a form in m variables, the new form $f(g|g|g|\cdots|g)$ has degree de and is a form in nm variables. If both f and g are normic of order i, so is $f(g|g|\cdots|g)$. In particular, if f is normic of order i, we may set $f = g$ in the above procedure and obtain the form $f^{(1)} = f(f|f|\cdots|f)$. By iterating the procedure we produce the sequence of normic forms of order i: $f^{(1)}, f^{(2)}, \ldots, f^{(\ell)}$,

$$f^{(\ell)} = f^{(\ell-1)}(f|f|\cdots|f) .$$

Observe that the degree of $f^{(\ell)}$ is $d^{\ell+1}$, where d is the degree of f. Since the norm form of a finite extension is normic of order 1, *every non-algebraically closed field* k *possesses normic forms* (= normic of order 1) *of arbitrarily large degree.*

LEMMA 7. (Artin, Lang, Nagata). *Let* k *be a* C_r *field, and let* $f_1, \ldots, f_s$ *be forms in* n *common variables each of degree* d. *If* $n > sd^r$, *then these forms have a non-trivial common zero in* k.

Proof: If k is algebraically closed $(r = 0)$, then the ordinary intersection dimension theorem of projective space yields the result. Hence, we may and do assume that $r > 0$; so there exists a normic form, N, of very large degree (as yet unspecified), say t. Consider the forms

$$\Phi^{(\ell)} = \Phi^{(\ell-1)}(f_1, \ldots, f_s | f_1, \ldots, f_s |, \cdots, | f_1, \ldots, f_s |, 0, \ldots, 0) ,$$

$$\Phi^{(1)} = N(f_1, \ldots, f_s | f_1, \ldots, f_s |, \ldots, | f_1, \ldots, f_s |, 0, \ldots, 0) ,$$

where we insert as many batches of $f_1, \ldots, f_s$ as possible, and we use new variables after each vertical slash. By the division algorithm, the number of zeros in each of the $\Phi^{(\ell)}$ is less than s; for if v_ℓ is the number of zeros in $\Phi^{(\ell)}$ and if n_ℓ is the number of variables of $\Phi^{(\ell)}$, then $n_\ell = n \cdot m_\ell$, and

$$(*) \qquad s\,m_\ell + v_\ell = n_{\ell-1}\,, \qquad 0 \leqq v_\ell < s\ .$$

(Of course, $n_0 = t$.) The degree of $\Phi^{(\ell)}$ is $t_\ell = t\,d^\ell$.

If we can arrange matters so that $n_\ell > (t_\ell)^r$ for some suitable ℓ, then our hypothesis will yield a non-trivial zero of $\Phi^{(\ell)}$. Since N is a normic form, this will give the simultaneous vanishing of $f_1, \ldots, f_s$ on some non-trivial point and complete the proof.

To prove that $n_\ell > (t_\ell)^r$, we need only unwind the recursive relations (*) connecting the n_ℓ and use our freedom in choosing a large value of t. The details are somewhat messy, they go as follows: Introduce the notations

$$q = \frac{n}{s\,d^r} > 1 \quad \text{(by hypothesis)},$$

$$\xi_{\ell-j} = \frac{s}{t^r (d^r)^{\ell-j}}\,, \qquad \xi = \xi_{\ell-1}\ .$$

Observe that $q^{j-1}\xi_{\ell-j} = \left(\frac{n}{s}\right)^{j-1}\xi$.

Now $n_\ell = n\,m_\ell$, and $n_{\ell-1} = s\,m_\ell + v_\ell$ with $0 \leqq v_\ell < s$. Thus,

$$\frac{n_\ell}{(t_\ell)^r} = \frac{n\,m_\ell}{t^r d^{\ell r}} = \frac{n\left(\frac{n_{\ell-1}}{s}\right) - n\left(\frac{v_\ell}{s}\right)}{t^r d^{\ell r}}\ ;$$

so,

$$(**) \qquad \frac{n_\ell}{(t_\ell)^r} > q\left(\frac{n_{\ell-1}}{(t_{\ell-1})^r} - \xi_{\ell-1}\right)\ .$$

We apply (**) recursively for $\ell-1, \ell-2, \ldots, 3, 2,$ and we get

$$\frac{n_\ell}{(t_\ell)^r} > q^2 \frac{n_{\ell-2}}{(t_{\ell-2})^r} - q\xi((\tfrac{n}{s}) + 1)$$

$$\frac{n_\ell}{(t_\ell)^r} > q^3 \frac{n_{\ell-3}}{(t_{\ell-3})^r} - q\xi((\tfrac{n}{s})^2 + (\tfrac{n}{s}) + 1)$$

$$\vdots$$

$$\frac{n_\ell}{(t_\ell)^r} > q^{\ell-1} \frac{n_1}{(t_1)^r} - q\xi \sum_{j=0}^{\ell-2} (\tfrac{n}{s})^j .$$

This yields the inequality

$$\frac{n_\ell}{(t_\ell)^r} > q^{\ell-1} \frac{n_1}{(t_1)^r} - q\xi \frac{(\frac{n}{s})^{\ell-1} - 1}{(\frac{n}{s}) - 1}, \quad \text{or}$$

$$\frac{n_\ell}{(t_\ell)^r} > q^{\ell-1} \frac{n_1}{(t_1)^r} - \frac{q\xi s}{s^{\ell-1}} \left(\frac{n^{\ell-1} - s^{\ell-1}}{n - s} \right) .$$

However, $t_1 = td$; $n_1 = nm_1$ and $t = n_0 = sm_1 + v_0$. Hence,

$$\frac{n_\ell}{(t_\ell)^r} > q^\ell \left(\frac{t - v_0}{t^r} \right) - \frac{q\xi s}{n-s} \left(\frac{n^{\ell-1} - s^{\ell-1}}{s^{\ell-1}} \right) .$$

Substituting for ξ, and simplifying, we get

$$(***) \qquad \frac{n_\ell}{(t_\ell)^r} > q^\ell \left[\frac{(t - v_0)(n - s) - s^2}{t^r} \right] + \frac{1}{(d^r)^{\ell-1}} \left(\frac{qs^2}{t^r(n-s)} \right) .$$

Choose t so large that $(t - v_0)(n - s) - s^2$ is positive. As $q > 1$, this will insure that as $\ell \to \infty$ the first term on the right in (***) increases without bound, while the second term shrinks to zero because its denominator $(d^r)^{\ell-1} \to \infty$ (as $d > 1$). Our proof is complete. Q.E.D.

PROPOSITION 33. *Let* k *be a field and let* K/k *be an algebraic extension.*

(1) *If* k *is* C_r *so is* K.

(2) *If* k *is* QAC *(i.e., has* C_1*), and if* L/K *is a finite extension, then* $\mathfrak{N}_{L/K} L^* = K^*$.

Proof: (1) Let f be a form of degree d in n variables over K. Assume $n > d^r$. The coefficients of f lie in a finite extension of k, so we may assume K/k is finite. Let $e_1, \ldots, e_s$ be a k-basis for K, introduce indeterminates X_{ij}, $i = 1, 2, \ldots, n$; $j = 1, \ldots, s$, and set

$$Y_i = \sum_{j=1}^{s} X_{ij} e_j .$$

Then

$$f(Y_1, \ldots, Y_n) = f_1(X_{ij}) e_1 + \cdots + f_s(X_{ij}) e_s ,$$

where each f_i is a form over k of degree d in ns variables. Since $ns > sd^r$ (because $n > d^r$), Lemma 7 implies the f_ν have a common non-trivial zero. That is, f has a non-trivial zero in K.

(2) According to part (1) of the proof, we may assume K = k, and thus L/k is finite. Let $a \in k^*$ and consider the equation $\mathfrak{N}_{L/k}(x) = a x_0^d$, where $d = [L:k]$. If we write $x = \Sigma x_i e_i$, where e_i is a k-base for L, we see that $\mathfrak{N}_{L/k}(x) - a x_0^d$ is a form in d+1 variables $x_0, x_1, \ldots, x_d$ of degree d. By hypothesis it has a *non-trivial* zero. Were all the x_i zero, it would follow that x_0 is also zero, a contradiction. Hence, at least one of the x_i is not zero; therefore, $\mathfrak{N}_{L/k}(x) \neq 0$, so that $x_0 \neq 0$. This yields the equation

$$\mathfrak{N}_{L/k}\left(\frac{x}{x_0}\right) = a$$

for some choice $x = \langle \beta_1, \ldots, \beta_n \rangle$, $x_0 = \beta_0$ of elements of k. Q.E.D.

COROLLARY 1. *A* QAC *field has* dim $\leqq 1$. *Therefore a* QAC *field has a trivial Brauer group along with all its algebraic extensions.*

Remark. Corollary 1 states that a QAC field possesses no non-trivial finite division algebras over it. This again shows how the notion C_1 generalizes the notion of algebraic closure.

COROLLARY 2. *Suppose* k *is a QAC field of characteristic* $p > 0$. *Then either* k *is perfect or* $[k : k^p] = p$. *In the latter case, the only purely inseparable extensions of* k *are the fields* $k^{p^{-i}}$ *for* $i = 0, 1, 2, \ldots, \infty$.

Proof: Suppose $k \neq k^p$. Let K/k be a purely inseparable extension of degree p, then Proposition 33 shows that $\mathfrak{N}_{K/k}K^* = k^*$. It follows that $K^p = k$; thus, $K^{p^2} = k^p$. But,

$$[k : k^p] = [K^p : K^{p^2}] = [K : K^p] = [K : k] = p\ ,$$

and we also have $K = k^{1/p}$. Q.E.D.

There are three classical theorems showing that certain fields are QAC. The first of these–the famous "Tsen's Theorem" was the starting point of the whole theory as mentioned above.

THEOREM 24. (Tsen). *Let* k *be a function field of dimension* 1 *over an algebraically closed field* k_0. *Then* k *is QAC.*

Proof: (Lang). According to Proposition 33, we may assume $k = k_0(t)$ –the rational function field in one variable over k_0. Let $f(x_1, \ldots, x_n)$ be a form of degree d in more than d variables, having coefficients in $k_0(t)$. After clearing denominators, we may assume f has coefficients in $k_0[t]$. Pick a large and as yet unspecified integer s, and introduce new indeterminates ξ_{ji}, $i = 0, 1, \ldots, s$; $j = 1, \ldots, n$; then set

$$(*) \qquad x_j = \sum_{i=0}^{s} \xi_{ji} t^i\ .$$

The form f looks like $\Sigma\, a_{(j)} x^{(j)}$ where j is an n-tuple $\langle j_1, \ldots, j_n \rangle$ such that $j_1 + \cdots + j_n = d$; so if we substitute equation (*) for each x_j we obtain

$$f(x_1, \ldots, x_n) = f_0(\xi) + f_1(\xi)t + \cdots + f_{ds+r}(\xi)t^{ds+r} \quad .$$

Here, each $f_i(\xi)$ is a form of degree d in the $n(s+1)$ variables ξ_{ji}, the integer r comes from the fact that our coefficients for f are polynomials in t, and each f_i has coefficients in k_0. But the system of equations $f_0(\xi) = \cdots = f_{ds+r}(\xi) = 0$, for large s, defines a projective variety in projective $n(s+1)-1$ space. The intersection dimension theorem tells us that each irreducible component of this variety has dimension $\geqq$ $n(s+1) - (ds+r+1)$. As $n > d$, it follows that for large s, our projective variety has points rational over k_0. (Here is where we use the algebraic closedness of k_0 !) Such a point (ξ) yields a non-trivial zero for the simultaneous equations $f_0(\xi) = \cdots = f_{ds+r}(\xi) = 0$; hence, formula (*) gives us the required non-trivial zero for f in $k_0(t)$. Q.E.D.

COROLLARY (Original Tsen Theorem). *If* k *is a function field in one variable over an algebraically closed field, then the Brauer group of* k *is trivial.*

Remark. Lang's proof actually generalizes to yield a "transition theorem" for the property C_r. Namely, *Lang proved that if* k *is a function field in* s *variables over a* C_r *field* k_0, *then* k *has* C_{r+s}. We refer the reader to [LQ] for details.

On the strength of his reformulation of Tsen's Theorem, and a very well-known result of Wedderburn (see the Corollary below), Artin conjectured the next theorem. It was proved almost immediately by Chevalley.

THEOREM 25 (Chevalley). *Every finite field is QAC.*

Proof: (J. Ax). We shall prove a stronger result. First, the form f need not really be homogeneous, all we need is a polynomial of degree d in more than d variables with *zero constant term* and we get a non-trivial zero. (This was shown by Chevalley.) Second, if f is such a polynomial, let $N(f)$ denote the number of zeros of f in k *including the trivial zero,* we shall show that the characteristic, p, of k divides $N(f)$. (This was proved by Warning [Wg].)

Each element of k is a $q-1^{st}$ root of unity, or it is zero—since we assume the cardinal of k is q. Let x denote an n-tuple of elements of k, where n is the number of variables in f. Then $f(x) \in k$, so $1-f(x)^{q-1}$ is 0 if $f(x) \neq 0$, and is 1 otherwise. So, for each zero of f(x) in k, $1-f(x)^{q-1}$ is 1, but makes no contribution for the other x. Therefore,

$$N(f) = \sum_{x \in k^n} 1-f(x)^{q-1} = \sum_{x \in k^n} 1- \sum_{x \in k^n} f(x)^{q-1} \quad ;$$

hence,

$$N(f) \equiv - \sum_{x \in k^n} f(x)^{q-1}, \quad (\text{mod } p) \ .$$

The polynomial $f(x)^{q-1}$ has degree $(q-1)d$; so it is a linear combination of monomials (over k) of degrees at most $(q-1)d$. Choose such a monomial, say $x_1^{u_1} x_2^{u_2} \dots x_n^{u_n}$ with $\Sigma\, u_i \leqq (q-1)d$, and let us compute the sum of the values of this monomial over all n-tuples from k. We get,

$$\sum_{x \in k^n} (x_1^{u_1} x_2^{u_2} \dots x_n^{u_n}) = \sum_{x \in k^n} \prod_{j=1}^{n} x_j^{u_j} = \prod_{j=1}^{n} \sum_{x_j \in k} x_j^{u_j} \ .$$

Now $\Sigma\, x_j^{u_j}$ is a sum of $(q-1)^{st}$ roots of unity (the term $x_j = 0$ contributes nothing); hence it vanishes if u_j is not a positive multiple of $q-1$. If u_j is a positive multiple of $q-1$, each term is 1, and the sum is $q-1$ which is -1 in the field k. Thus,

$$(**) \qquad \sum_{x \in k^n} (x_1^{u_1} \dots x_n^{u_n}) = \prod_{j=1}^{n} Y(u_j) \ ,$$

where $Y(u_j) = -1$ iff u_j is a multiple of $q-1$, zero otherwise. Now we know $\Sigma\, u_i \leqq d(q-1) < n(q-1)$, so at least one u_j *cannot be a multiple of* $q-1$; and for this u_j the term $Y(u_j)$ vanishes. Equation (**) shows that each monomial sums to zero over k^n; hence, every polynomial of required degree also sums to zero. Thus,

$$N(f) \equiv - \sum_{x \in k^n} f(x)^{q-1} = 0 \quad (\text{mod } p). \qquad \text{Q.E.D.}$$

COROLLARY (Wedderburn). *Every finite division ring is commutative.*

Proof: For this statement we can give two proofs. First, by Corollary 1 of Proposition 33 and Theorem 25, the Brauer group of a finite field is trivial. But (as in the remark following Corollary 1 of Proposition 33) the Brauer group of a field classifies the equivalence classes of central simple k-algebras (i.e., algebras simple over k with k as center), and in each class there is a unique division ring with center k. The classes are trivial iff their corresponding division rings are just the field k. So given a division ring D, its center k is a finite field; and $\mathrm{Br}(k) = (0)$ implies D equals its center.

For a second proof, let D be a division ring, let k be its center. One knows that $[D:k] = r^2$ for some integer r, so the reduced norm: $D^* \to k^*$, say $\mathfrak{N}_0$ yields a form in r^2 variables of degree r *via*

$$f(x_1, \ldots, x_{r^2}) = \mathfrak{N}_0(\sum_j x_j e_j) \quad ,$$

where the e_j are a k-basis for D. However, C_1 implies this form has a non-trivial zero, a contradiction. Therefore, $r = 1$. Q.E.D.

The third classical result is Lang's theorem that a complete, discretely valued field with algebraically closed residue field has property C_1. The rest of this section will be devoted to an exposition of this important theorem.

LEMMA 8 (Newton Approximation Lemma). *Let* k *be complete with respect to the non-archimedian valuation* ord_k *(exponential form). Let* $\mathcal{O}$ *be the valuation ring in* k, *and let* $f(x) \in \mathcal{O}[x]$. *If* $a_0 \in \mathcal{O}$, *and if* $\mathrm{ord}_k(f(a_0)/f'(a_0)^2) > 0$, *then the sequence*

$$a_{i+1} = a_i - \frac{f(a_i)}{f'(a_i)}$$

converges to a root a *of* $f(x)$. *Moreover,* $\mathrm{ord}_k(a - a_0) \geqq \mathrm{ord}_k(\frac{f(a_0)}{f'(a_0)})$.

We shall leave the proof to the reader, it is a simple exercise in non-archimedian calculus.

We wish to observe, however, that the hypotheses of Lemma 7 are *open conditions*. So, if $f(X_1, \dots, X_n)$ is a polynomial in n variables and $\alpha = \langle \alpha_1, \dots, \alpha_n \rangle$ has the property that $f(X_1, \alpha_2, \dots, \alpha_n)$ satisfies the hypotheses of Lemma 7 with respect to α_1 (i.e., α_1 may be refined to a root of $f(X, \alpha_2, \dots, \alpha_n)$), then when $\beta = \langle \beta_1, \dots, \beta_n \rangle$ is sufficiently close to $\langle \alpha_1, \dots, \alpha_n \rangle$, we deduce that β_1 may be refined to a root of $f(X, \beta_2, \dots, \beta_n)$.

Now let k denote a complete, discretely valued field with algebraically closed residue field k_0. Then one knows that k is a purely ramified extension of an invariantly determined subfield k_u, itself complete, discretely valued and with residue field k_0. In the equi-characteristic case, k_u is k itself; in the unequal characteristic case, k_u is the fraction field of the Witt vectors on k_0. (See for example Serre's book [SCL], or Mumford's notes [ML], or Witt's paper itself [W]—a classic paper.) In any case, each element $x \in \mathcal{O}_u$—the ring of integers of k_u—is a vector $\langle \xi_0, \xi_1, \dots \rangle$ where $x = \Sigma \xi_j t^j$ in the equal characteristic case or $x = \langle \xi_0, \dots \rangle$ as Witt vector. Here, the ξ_j belong to k_0.

If $f(X_1, \dots, X_n)$ is a polynomial with coefficients in $\mathcal{O} = \mathcal{O}_u$, and if $x_1, \dots, x_n \in \mathcal{O}$, then

$$f(x_1, \dots, x_n) = \langle f_0(\xi_0), f_1(\xi_0, \xi_1), \dots \rangle$$

where the polynomials f_j have coefficients in k_0. Consequently, to find a zero of f in $\mathcal{O}$ is to solve the infinite system $f_j = 0$ of polynomials in countably many variables for a solution in k_0. Let $A = k_0[\xi_0, \dots]$ be the ring of polynomials in countably many indeterminates. If $\mathfrak{A}$ is an ideal of A, let $\underline{m}$ be a maximal ideal of A which contains $\mathfrak{A}$. We wish to prove $V(\mathfrak{A}) \neq \emptyset$; as $\underline{m} \supseteq \mathfrak{A}$, we deduce $V(\underline{m}) \subseteq V(\mathfrak{A})$, so it suffices to prove $V(\underline{m}) \neq \emptyset$. The ring $A/\underline{m}$ has the form $k_0[\eta_0, \dots]$ where the η_j lie in a large field extension of k_0. Let k_1 be the algebraic closure of $k_0(\eta_0, \dots)$ and let K be the Witt field with residue class field k_1. Then when $\mathfrak{A} = (f)$ (i.e., $\mathfrak{A}$ = ideal of A corresponding to some $f \in \mathcal{O}[X_1, \dots, X_n]$), it is clear that f has a zero in K. This proves

LEMMA 9. *If* $f \in \mathcal{O}[X_1, \ldots, X_n]$, *there exists a field extension,* K, *of* k *which is complete and has an algebraically closed residue field, such that* f *has a zero in* K *provided* f *is not constant.*

PROPOSITION 34. *Let* K *and* k *be the fields of the above paragraph. Then, in the terminology of Weil* [W1], K/k *is a regular extension. That is,* k *is algebraically closed in* K, *and every finitely generated subextension of the layer* K/k *is separably generated.*

Proof: Let L be a finite subextension of K/k. Since k_0 is algebraically closed, L has the same residue field as k; hence, L is totally ramified over k. But K is unramified over k and $L \subseteq K$; therefore, $L = k$. For the second part, observe that the question can only arise in the equi-characteristic case. In this case, we have $k = k_0((t))$, $K = k_1((t))$. Now let L be a finitely generated extension in the layer K/k, then all we need to prove is that L is linearly disjoint from $k^{1/p}$ over k. The field $k^{1/p}$ is exactly $k_0((\theta))$ where $\theta^p = t$. Let $u_1, \ldots, u_n$ be linearly independent (over k) elements chosen from L. We must show that these elements remain linearly independent over $k_0((\theta)) = k_2$. Choose a linear relation $\Sigma\, c_j u_j = 0$ with coefficients $c_j \in k_2$. Write the c_j as power series in θ and the u_j as power series in t. (We may obviously assume the c_j are integral power series in θ, in fact we may even assume one of the c_j is a unit of $k_0[[\theta]]$.) Since $t = \theta^p$, all terms in the expanded power series having θ^n with $n \not\equiv 0$ (mod p) come from the c_j *not* from the u_j alone. The power series being equal to zero implies all its coefficients vanish, Hence, because these coefficients are linear homogeneous forms in the coefficients of the power series for c_j, we deduce that all coefficients of the various power series for the c_j of terms involving θ^n ($n \not\equiv 0 \pmod p$) vanish. It follows that each c_j is a power series in $\theta^p = t$; hence, each $c_j \in k$. Therefore $c_j = 0$ for all j by the linear independence of the u_j over k. Q.E.D.

THEOREM 26 (Lang's Specialization Theorem). *Let* k_0, k_1 *be algebraically closed fields and let* k, K *be the Witt fields over them. Let*

$(\eta) = \langle \eta_1, \ldots, \eta_n \rangle \in K^n$ *be a point NOT rational over* k, *and let* Ω *be a dense subfield of* k, *algebraically closed in* k. *Then there exists a specialization of* (η) *to a point rational over* Ω.

Proof: Let $L = k(\eta) = k(\eta_1, \ldots, \eta_n)$, then by Proposition 34 L/k is separably generated. Say $t_1, \ldots, t_r$ are a separating transcendence base for L/k, then we may write $L = k(t_1, \ldots, t_r\ \alpha)$ where α is separable algebraic over $k(t_1, \ldots, t_r)$. By multiplication by a suitable high power of the prime element of k, we may assume each t_j is an integral power series, and that α is an integral power series. Since each η_j lies in L, we can write

$$\eta_j = \Sigma_\nu \frac{\phi_{j\nu}(t)}{\psi_{j\nu}(t)} \alpha^\nu \quad ,$$

where the $\phi_{j\nu}$ and $\psi_{j\nu}$ are polynomials in t with coefficients in k. Now α satisfies an irreducible equation $g(y,t) = 0$ over $k(t_1, \ldots, t_r)$, and if we multiply through by all denominators, we may assume the coefficients of $g(y, t_1, \ldots, t_r)$ are *polynomials* in t.

Let V_η be the locus of (η) over k, and let $f_1, \ldots, f_\ell$ be a basis for the corresponding prime ideal in $k[X_1, \ldots, X_n]$. Thus, $f_j(\eta) = 0$ for all j. If y is an indeterminate, set

$$(*) \qquad f_j((\Sigma_\nu \frac{\phi_{i\nu}(t)}{\psi_{i\nu}(t)} y^\nu)) = g_j(y) \quad .$$

By our expression for η_j (for each j), we see that $g_j(\alpha) = 0$ for each j. It follows that $g_j(y)$ lies in the ideal generated by $g(y, t)$ in the ring $k(t_1, \ldots, t_r)[y]$. Consequently, there exist $h_j(y, t)$ with

$$f_j((\sum_\nu \frac{\phi_{i\nu}(t)}{\psi_{i\nu}(t)} y^\nu)) = g(y,t) h_j(y,t) \qquad \text{for } j = 1, \ldots, \ell \quad .$$

Expand each t_j and α in vectors (as explained in the discussion after Lemma 8), *viz*:

$$\left.\begin{array}{l} t_j = (t_{j0}, t_{j1}, \ldots) \\ \alpha = (\alpha_0, \alpha_1, \ldots) \end{array}\right\} \quad t_{ij}, \alpha_i \in k_1 \quad .$$

Choose a large but as yet undetermined integer s, and truncate these vectors after the sth stage. Examine the locus of the point

$$(\alpha_0, \alpha_1, \dots, \alpha_s, t_{10}, \dots, t_{r0}, \dots, t_{1s}, \dots, t_{rs})$$

over k_0. (Call this point $(\alpha_i, t_{ij})_s$.) Now $\partial g/\partial y\ (\alpha) \neq 0$ because α is separable over $k(t_1, \dots, t_r)$. Let us expand both $g(y, t)$ and $\partial g/\partial y$ as vectors and evaluate them on (α, t). We get

$$g(\alpha, t) = (g_0(\alpha_i, t_{ij}),\ g_1(\alpha_i, t_{ij}), \dots)$$
$$\frac{\partial g}{\partial y}(\alpha) = (g_0'(\alpha_i, t_{ij}),\ g_1'(\alpha_i, t_{ij}), \dots)\ ;$$

since $g(\alpha, t) = 0$, every component $g_\nu(\alpha_i, t_{ij})$ vanishes as well. But $\partial g/\partial y\,(\alpha) \neq 0$, so there is some component, say $g_m'(\alpha_i, t_{ij})$, which does not vanish. Let $(\beta_0, \dots, \beta_s, u_{10}, \dots, u_{r0}, \dots, u_{1s}, \dots, u_{rs})$ be some specialization of $(\alpha_i, t_{ij})_s$ *in* k_0, and construct the elements

$$u_1 = (u_{10}, u_{11}, \dots, u_{1s}, 0, 0, \dots)$$
$$u_2 = (u_{20}, \dots\dots, u_{2s}, 0, 0, \dots)$$
$$\vdots$$
$$u_r = (u_{r0}, \dots\dots, u_{rs}, 0, 0, \dots)$$
$$\beta = (\beta_0, \dots\dots, \beta_s, 0, 0, \dots)$$

of k.

In expression (*), there occur only finitely many denominators, each a polynomial in t. If we expand each denominator as a vector, then each will have a non-vanishing component, say of the form $\psi(t_{ij})$. Consider these finitely many non-vanishing components (one from each denominator vector) and the non-vanishing $g_m(\alpha_i, t_{ij})$. This gives us a finite family of polynomials *over* k_0 in the variables t_{ij}, α_i —none of which is zero. It follows that for large enough s, none of these polynomials vanish on the point $(\alpha_i, t_{ij})_s$ —so their product does not vanish. It follows as well (from the

Hilbert Nullstellensatz) that we may choose the above specialization: $(\alpha_i, t_{ij})_s \longmapsto (\beta_i, u_{ij})_s$ so that this product does *not* vanish on the point $(\beta_i, u_{ij})_s$. When we do this, if s is *very large,* we find that $g(\beta, u)$ is close to $g(\alpha, t) = 0$ by continuity, the denominators $\psi_{i\nu}(u)$ will not be zero, and neither will $g_m(\beta_i, u_{ij})$ be zero. Thus, $\partial g/\partial\beta$ is very large compared with $g(\beta, u)$; so Newton's Lemma applies.

If we now choose $w \in \Omega$ very close to u, continuity shows that the hypotheses of Newton's Lemma remain verified for $g(\beta, w)$ and $\partial g/\partial y(\beta)$. Thus β may be refined to a root, say γ, of $g(y, w)$. This γ is algebraic over Ω and lies in k. Hence, $\gamma \in \Omega$; and if we set

$$\bar{\eta}_j = \sum_\nu \frac{\phi_{j\nu}(w)}{\psi_{j\nu}(w)} \gamma^\nu , \text{ the point } (\bar{\eta}) = (\bar{\eta}_1, \dots, \bar{\eta}_n)$$

is the required specialization. Q.E.D.

THEOREM 27 (Lang). *Let* k *be a field complete with respect to a discrete valuation having algebraically closed residue field. Let* Ω *be a dense subfield of* k *such that* k/Ω *is a regular extension (*cf. Proposition 34*). Then* Ω *is a* C_1 *field. In particular,* k *is* QAC.

Proof: As usual, it suffices to assume k is the Witt field over the residue field k_0. Let $f(X_1, \dots, X_n)$ be a form of degree d in n variables with $d < n$ and coefficients in Ω. Since we may obviously assume the coefficients of f are integers of k, we are placed in the situation discussed prior to Lemma 9. Let $f_0, f_1, f_2, \dots,$ be the countable system of polynomials (over k_0) in countably many indeterminates corresponding to the form f. We shall show that this system has a non-trivial zero—or, what is the same, that the ideal $\mathfrak{A}$ generated by the f_j (for all j) is not the unit ideal. When this is done, Theorem 26 will yield the required non-trivial zero in Ω.

Consider the partial system $f_0, \dots, f_s$. Each f_j is a form in the variables $\xi_{10}, \dots, \xi_{n0}, \xi_{11}, \dots, \xi_{n1}, \dots, \xi_{1j}, \dots, \xi_{nj}$; hence, the forms $f_0, \dots, f_s$ define a variety V_s in the space of the variables $\xi_{10}, \dots, \xi_{n0}, \dots, \xi_{1r}, \dots, \xi_{ns},$

that is, in $A^{(s+1)n}$ space. Consider affine n-space A^n as represented by the variables $\xi_{10}, \ldots, \xi_{n0}$ and consider the projection of V_s on A^n, call it, $\mathrm{pr}(V_s)$. The crux of the argument will be to show that for every s, $\mathrm{pr}(V_s)$ does not consist only of the trivial zero $\langle 0, 0, \ldots, 0 \rangle$. Indeed, let us assume this statement for each s and show how the argument is completed from it. Our statement implies that if is the valuation ideal of the ring, $\mathcal{O}$, of integers in k, then the form $f(X_1, \ldots, X_n)$ has solutions mod $\mathfrak{p}^\nu$ for every ν, and *that in these solutions at least one* x_i *is a unit* (because $\xi_{i0} \neq 0$ for some i by assumption). By homogeneity, we may assume one of the x_i is actually 1; so we have shown that for every ν, there is an integer $i(\nu)$, $1 \leqq i(\nu) \leqq n$, such that

$$f(X_1, \ldots, X_{i(\nu)-1}, 1, X_{i(\nu)+1}, \ldots, X_n) \equiv 0 \pmod{\mathfrak{p}^\nu} .$$

is solvable in k (in fact in $\mathcal{O}$). Since there are infinitely many ν, at least one index i works for an infinite set of ν, say $i = 1$. Therefore, we now deduce that we may solve $f(1, X_2, \ldots, X_n) \equiv 0 \pmod{\mathfrak{p}^\nu}$ for every ν; hence, the ideal generated by $f_0(1, \ldots)$, $f_1(1, \ldots)$, $f_2(1, \ldots)$, ..., $f_r(1, \ldots)$, ... cannot be the unit ideal (else the unit element would be a finite combination of the f_j and for some ν we would not be able to solve our congruence). This is exactly what we set out to prove, so our claim is established.

It remains, therefore, to prove $\mathrm{pr}(V_s) \neq \langle 0, 0, \ldots, 0 \rangle$ for each s. Denote the n-tuples $\langle \xi_{1j}, \ldots, \xi_{nj} \rangle$ by the single letter ξ_j, so that in this notation, each f_j is a form in the variables $\xi_0, \xi_1, \ldots, \xi_j$. Suppose for some s that the variety V_s has only points for which $\xi_0 = 0$, or what is the same, that the only solutions to the congruence $f(X_1, \ldots, X_n) \equiv 0 \pmod{\mathfrak{p}^s}$ are those vectors X_j whose zeroth components vanish. In the unequal characteristic case we have the relation $p = FV = VF$, where F is the Frobenius map and V is the *Verschiebung* (the shift one to the right). Our vectors X_j have the form $\langle 0, \xi_{j1}, \xi_{j2}, \ldots \rangle$; so if we set

$$Y_j = \langle \xi_{j1}^{1/p}, \xi_{j2}^{1/p}, \ldots \rangle$$

then $pY_j = X_j$. Hence, $f(X_1, \ldots, X_n) = p^d f(Y_1, \ldots, Y_n)$, which implies

$$<f_0(\xi_0), f_1(\xi_0, \xi_1), \ldots, f_m(\xi_0, \ldots, \xi_m), \ldots> =$$
$$p^d <f_0(\xi_1^{1/p}), \ldots, f_m(\xi_1^{1/p}, \ldots, \xi_{m+1}^{1/p}), \ldots> .$$

Now $p^d = V^d F^d$, so the last equation tells us that

(a) $f_r(0, \xi_1, \ldots, \xi_r) = 0 \quad \text{for } r < d$

(b) $f_r(0, \xi_1, \ldots, \xi_r) = f_{r-d}^{[p^d]}(\xi_1^{p^{d-1}}, \ldots, \xi_{r-d+1}^{p^{d-1}}), \quad \text{for } r \geqq d$

where $f_j^{[p^d]}$ means f_j with all its coefficients raised to the p^d-power. In the equal characteristic case, equations (a), (b) continue to hold except that the Frobenius doesn't arise so we may suppress the exponents p^d and p^{d-1}.

Now our assumption that $\mathrm{pr}(V_s) = <0, \ldots, 0>$ shows that $\mathrm{pr}(V_t) = <0, \ldots, 0>$ for any $t \geqq s$. In particular when $t = s+d$ the only solutions to $f_0 = \cdots = f_{s+d} = 0$ are the solutions of

$$f_0(0) = 0, f_1(0, \xi_1) = 0, \ldots, f_{s+d}(0, \xi_1, \ldots, \xi_{s+d}) = 0 .$$

By (a) and (b), these solutions are also the solutions of

$$f_0(\xi_1^{\alpha}) = f_1(\xi_1^{\alpha}, \xi_2^{\alpha}) = \cdots = f_s(\xi_1^{\alpha}, \ldots, \xi_{s+1}^{\alpha}) = 0$$

for a suitable exponent α. But $\mathrm{pr}(V_s) = <0, \ldots, 0>$; hence $\xi_1^{\alpha} = 0$, so we deduce $\xi_1 = 0$. By repeating the above argument or using induction, we see that for each integer $\ell \geqq 0$, the system

$$f_0 = f_1 = \cdots = f_{s+\ell d} = 0$$

has as solutions tuples $<\xi_0, \ldots, \xi_{s+\ell d}>$ *in which* $\xi_0, \ldots, \xi_\ell$ *all vanish.* The variety of zeros is embedded in affine $n(s+\ell d+1)$ space, and hence its dimension is $\leqq n(s+\ell d+1) - n(\ell+1) = n(s+(d-1)\ell)$. On the other hand, by the intersection dimension theorem, this variety has dimension $\geqq n(s+\ell d+1) - (s+\ell d+1) = (n-1)(s+\ell d+1)$. We deduce the inequality

$$(n-1)(s+\ell d+1) \leqq n(s+(d-1)\ell), \qquad \text{for all } \ell \geqq 0.$$

This yields the inequality

$$(n-d)\ell \leqq s+1-n, \qquad \text{for all } \ell \geqq 0.$$

But $n > d$, so for sufficiently large ℓ we get a contradiction; and the theorem is proved. Q.E.D.

§4. Existence of fields of arbitrary cohomological dimension

Our first observation is that there exist fields with infinite cohomological dimension. Indeed the real field, $\mathbf{R}$, is such a field for its Galois group is finite. A more difficult question is the existence of fields (hence, of profinite groups) with arbitrary finite cohomological dimension. The relevant theorem here was first conjectured by Grothendieck and proved by Tate. It was, in effect, the motivation for much of the theory of the cohomology of profinite groups.

THEOREM 28 (Tate). *Let* k' *be an extension of the field* k *of transcendence degree* n *(over* k*), and let* p *be a prime number. Then*

$$\mathrm{cd}_p\, k' \leqq n + \mathrm{cd}_p\, k$$

with equality if k' *is a function field over* k, $p \neq \mathrm{ch}\, k$, *and* $\mathrm{cd}_p\, k < \infty$.

Proof: k' is algebraic over a purely transcendental extension of k. As cohomological dimension decreases under algebraic extensions, it suffices to give the proof when k' is a pure transcendental extension over k. Moreover a simple induction shows that it suffices to give the proof when $n = 1$.

Let $\bar{k}$ be the algebraic closure of k, let $k' = k(t)$, and let $\overline{k(t)}$ be the algebraic closure of $k(t)$. Since $\bar{k}$ and $k(t)$ are linearly disjoint over k, we obtain the diagram of fields

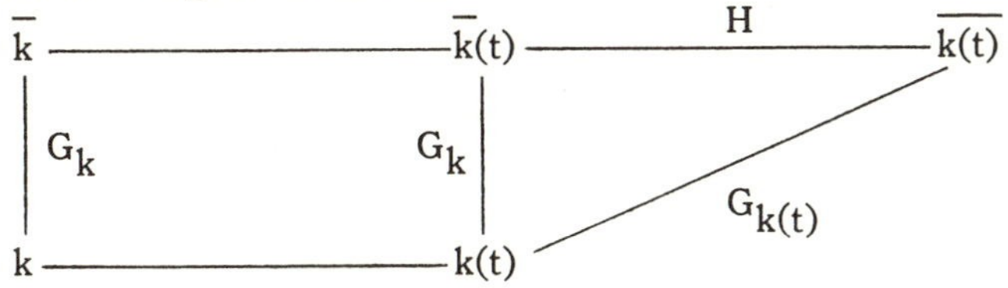

By the theorem on natural irrationalities, $G_k = G(\bar{k}(t)/k(t))$, and the letters H, $G_{k(t)}$ represent the Galois groups of the extensions as shown. Since $\bar{k}(t)$ is a function field with algebraically closed residue field, Tsen's theorem tells us that $\operatorname{cd} H \leqq 1$. But G_k is isomorphic to $G_{k(t)}/H$, so the Tower Theorem yields $\operatorname{cd}_p(k') \leqq \operatorname{cd}_p H + \operatorname{cd}_p(k) \leqq 1 + \operatorname{cd}_p(k)$. This proves the inequality.

For the equality, first note that the cohomological dimension of a finite extension is the same as that of the ground field. The obvious induction yields the fact that we need prove the result only in the case $k' = k(t)$, a pure transcendental extension of transcendence degree one. Moreover, by replacing k by one of its p-Sylow fields (where p is the given prime of the hypothesis), we may even assume G_k is a p-group. It will then follow that μ_p—the group of pth roots of unity in k—is actually isomorphic to Z/pZ as G_k-module and also as $G_{k(t)}$-module (because $\mu_p \subseteq k \subseteq k(t)$!). If $d = \operatorname{cd}_p k$, then the tower theorem and Tsen's theorem yield

$$H^{d+1}(k(t), \mu_p) = H^d(k, H^1(\bar{k}(t), \mu_p)) .$$

Now $H^1(\bar{k}(t), \mu_p) = \bar{k}(t)^*/\bar{k}(t)^{*p}$, so we deduce

$$(*) \qquad H^{d+1}(k(t), \mu_p) = H^d(k, \bar{k}(t)^*/\bar{k}(t)^{*p}) .$$

But there is the exact sequence

$$0 \longrightarrow \bar{U} \longrightarrow \bar{k}(t)^* \xrightarrow{\operatorname{ord}_{\bar{k}(t)}} Z \longrightarrow 0$$

arising from the obvious valuation $\operatorname{ord}_{\bar{k}(t)}$ of $\bar{k}(t)$. This yields the exact sequence

$$0 \longrightarrow \ker \longrightarrow \bar{k}(t)^*/\bar{k}(t)^{*p} \longrightarrow Z/pZ \longrightarrow 0 \ ;$$

and when we apply cohomology and use $\operatorname{cd}_p k = d$, we get the surjection $H^d(k, \bar{k}(t)^*/\bar{k}(t)^{*p}) \to H^d(k, Z/pZ) \to 0$. The right hand group does not vanish by Theorem 12, and equation (*) completes the proof. Q.E.D.

COROLLARY 1. *For every natural integer* $n \geqq 0$, *there exist fields, hence profinite groups, of cohomological dimension* n.

COROLLARY 2. *We have* $\mathrm{cd}\, k = 2$ *in each of the following cases:*

(a) $k = k_0(t)$; k_0 *finite,* k *a function field over* k_0 *of* tr.d. 1 *(case of Class Field Theory),*

(b) $k = k_0(x, y)$, k_0 *algebraically closed,* k *a function field in two variables over* k_0.

§5. Counter-examples in the theory of cohomological dimension and forms

We have shown that algebraic extensions of C_r fields are C_r and that the same transition property is valid for the property $\mathrm{cd}\, k \leqq r$. Moreover, the behavior of these two properties under transcendental extensions is the same and for perfect fields we know that C_1 implies $\mathrm{cd}\, k \leqq 1$. Therefore, the natural question arises: *What is the relation between* $\mathrm{cd}\, k \leqq r$ *and* C_r?

One *highly doubtful* hope was that $\mathrm{cd}\, k \leqq r$ implies k has C_r, but this was dashed by a beautiful counter-example of J. Ax which we shall treat below. The original hope stemmed from Artin's conjecture that p-adic fields were C_2. Indeed it is easy to see (cf. next chapter) that p-adic fields have cohomological dimension 2 (they even have strict cohomological dimension 2), so the hope was that this would lead to a proof of Artin's conjecture. In the equal characteristic case, Lang had proved Artin's conjecture, and one knew, moreover, that $\mathrm{cd}\, k \leqq 2$. But in 1966, G. Tirjanian [TJ] gave the first counter-example to Artin's conjecture for 2-adic fields and later that year S. Schanuel (unpublished) gave a whole series of counter-examples, one for each prime p. This development gave great weight to a famous theorem proved in 1964 by Ax-Kochen [AK] partially solving the Artin conjecture—we now see it as the best possible qualitative result of its type.

THEOREM 29 (Ax-Kochen). *For every positive integer* d, *there exists a finite set of primes,* $A(d)$, *such that for every prime* $p \notin A$, *every form* f

of degree d *in* $n > d^2$ *variables defined over* Q_p *(the rational* p*-adic field) has a non-trivial zero in* Q_p.

We refer the reader to [AK] for the proof of this extremely beautiful theorem. Right now we want to turn to Ax's counter-example on the implication "$\text{cd}\, k \leqq r \Longrightarrow k$ has C_r."

Let k_0 be a field of characteristic zero containing all the nth roots of unity for some $n > 0$. Let k be the field $k_0((t))(\{t^{1/m} \mid (m,n) = 1\})$, i.e., the composite of all field extensions of $k_0((t))$ generated by m^{th} roots of t for all positive m which are prime to n. It is very simple to see that the Galois group, G_k, of k satisfies

$$G_k = G_{k_0} \times \prod_{p \mid n} Z_p$$

where Z_p is the group of p-adic integers, i.e., proj lim Z/p^nZ. If we set $Z_{(n)}$ to be Z "localized at (n)," that is, the set of all rational numbers ξ/η such that $(\eta, n) = 1$, then $Z_{(n)}$ is an additive group and the valuation ord: $k_0((t)) \to Z \cup \{\infty\}$ (*via*: $t \to 1$) extends to a valuation

$$\text{ord}_k: \ k \ \to Z_{(n)} \cup \{\infty\},$$

with k_0 as residue class field.

LEMMA 10. *If* $H(U_1, \ldots, U_r)$ *is a form of degree* d *over* k_0 *with no non-trivial zero in* k_0, *then* H *has no non-trivial zero in* k.

Proof: For given elements $b_1, \ldots, b_r$ of k, not all zero, let us compute $\text{ord}_k H(b_1, \ldots, b_r)$. We choose $b \in k$ with $\text{ord}_k(b) = \inf_i \text{ord}_k(b_i)$ and set $a_i = b_i/b$, so that $\text{ord}_k(a_i) \geqq 0$ for all i and $\text{ord}_k(a_i) = 0$ for some i. Then

$$\begin{aligned}\text{ord}_k H(b_1, \ldots, b_r) &= \text{ord}_k(b^d H(a_1, \ldots, a_r)) \\ &= d\, \text{ord}_k(b) + \text{ord}_k H(a_1, \ldots, a_r)\ .\end{aligned}$$

Now if a bar on top represents passage to the residue class field, then $\overline{H(a_1, \ldots, a_r)} = H(\bar{a}_1, \ldots, \bar{a}_r)$. But, $\bar{a}_j \neq 0$ for some j by construction, so by hypothesis $H(\bar{a}_1, \ldots, \bar{a}_r) \neq 0$. It follows that $H(a_1, \ldots, a_r)$ is a unit of k; hence, its ordinal is zero. Therefore,

$$\mathrm{ord}_k H(b_1, \ldots, b_r) = d\ \mathrm{ord}_k(b) \neq \infty,$$

and this completes the proof. Q.E.D.

PROPOSITION 35. *There is a field* L *of (algebraic) dimension one, but* L *is not* C_1.

Proof: Choose k_0 algebraically closed and of characteristic zero, let $k = k_0((s))(s^{1/m}, (m, 6) = 1)$, and let $L = k((t))(t^{1/m}, (m, 5) = 1)$. By our remarks above,

$$G_L = \mathbf{Z}_2 \times \mathbf{Z}_3 \times \mathbf{Z}_5 \ ;$$

hence, $\mathrm{cd}_p L \leqq 1$ for all p, and it follows that $\dim L = 1$.

We shall construct a form of degree 5 in 10 variables with no non-trivial zero in L; this will show that L is not C_1. Let $J(X, Y)$ be a form of degree 2 over k with no non-trivial zero in k (for example, the norm form from $k(s^{1/2})$ to k) and let $K(X, Y, Z)$ be a form of degree 3 over k with no non-trivial zero in k (e.g., the norm form from $k(s^{1/3})$ to k). Set $M(X, Y) = J(X, Y) K(X, Y, 0)$, then $M(X, Y)$ is a form of degree 5 in two variables with no non-trivial zeros in k. From Lemma 10, we conclude that $M(X, Y)$ has no non-trivial zeros in L either; in fact, from the proof of Lemma 10, one sees that

$$\mathrm{ord}_L M(X, Y) \in 5\mathbf{Z}_{(5)} \quad \text{for all } \langle X, Y \rangle \in L^2 - \langle 0, 0 \rangle \ .$$

(Here, ord_L is the valuation on L having k as residue field.)

Set

$$H(Z_1, \ldots, Z_{10}) = M(Z_1, Z_2) + tM(Z_3, Z_4) + t^2 M(Z_5, Z_6) + t^3 M(Z_7, Z_8) + t^4 M(Z_9, Z_{10}) \ .$$

We contend H is the required form of degree 5 in 10 variables. To see this, observe that $H(Z_1, \ldots, Z_{10})$ is the sum of five terms at least one of which is non-zero by the above remarks. However, all these terms have ordinals different in $\mathbf{Z}_{(5)}$, even different modulo $5\mathbf{Z}_{(5)}$ by our construction; hence no cancellation can occur and $H(Z_1, \ldots, Z_{10}) \neq 0$. Q.E.D.

Remark: The field L of Proposition 35 is actually C_2 as one sees by examining the proof of Proposition 35.

To construct Ax's counter-example we proceed as follows: Start with k_0 as in Proposition 35, and define k_p inductively for ascending prime numbers p by:

$$k_2 = k_0((t_2))(t_2^{1/n};\ (n, 2) = 1)$$

if p is given and q is the largest prime $< p$, then

$$k_p = k_q((t_p))(t_p^{1/n};\ (n, p) = 1)\ .$$

Now let $R = \operatorname{dir\,lim}_p k_p$, then R is a field, and we see from previous remarks that $G_R = \Pi_p Z_p = \hat{Z}$. Hence, R is *quasi-finite, and* $\dim R = 1$.

THEOREM 30 (Ax's [Ax]). *The field* R *constructed above has* $cd_p R \leqq 1$ *for all* p, *but* R *is NOT* C_r *for any* r.

Proof: Choose r, we shall show that R is not C_r. If n is an integer $> r$, choose some α with $0 < \alpha < 1$ such that

$$\lambda = \lambda(n, \alpha) = \frac{1-\alpha^n}{1-\alpha} = \sum_{i=0}^{n-1} \alpha^i > r\ .$$

Given an integer $m \geqq 0$, let us call a prime number p "m-representable" iff either $m = 0$ or $p = p_1 + p_2 + p_3$ for prime numbers p_1, p_2, p_3 which are $m-1$ representable and $p^\alpha < p_1 < p_2 < p_3$. We consider the following two statements about m-representable prime numbers:

(*) *If* p *is* m-*representable, there exists a form of degree* p *in at least* $p^{\lambda(m,\alpha)}$ *variables having no non-trivial zero in* k_p, and

(**) *Given* m, *there is an integer* c *such that every prime* $p > c$ *is* m-*representable.*

From (*) and (**), we deduce the existence of a form of degree p in at least $p^\lambda > p^r$ variables ($\lambda = \lambda(n, a)$) having no non-trivial zero in k_p. By Lemma 10, such a form has no non-trivial zero in R; and it follows that R is not C_r. Consequently, it remains to prove (*) and (**).

We prove (*) by induction on m. For $m = 0$, $\lambda(0, a)$ does not exist so (*) is vacuously true. Therefore, we may assume $m \geqq 1$ and that (*) is true for $m-1$. The prime p is the sum $p_1 + p_2 + p_3$, with $p^a < p_1 < p_2 < p_3$, and each p_i is $(m-1)$ representable. Let $\mu = \lambda(m-1, a) = \Sigma_{i=0}^{m-2} a^i$, then there exist forms $H_i(U_1, \ldots, U_{v_i})$ of degrees p_i over k_{p_i} having no non-trivial zeros in k_{p_i}. Here,

$$v_i \geqq p_i^\mu \geqq p_1^\mu .$$

By setting the extra variables equal to zero, we may assume $v_i = v_1 = v$, and we know $v \geqq p_1^\mu$. Let

$$J(U_1, \ldots, U_v) = \prod_{i=1}^{3} H_i(U_1, \ldots, U_v) \in k_{p_3}[\cdots] ,$$

and consider

$$H(Z_1, \ldots, Z_{pv}) = \sum_{i=0}^{p-1} t_p^{\,i} J(Z_{iv+1}, \ldots, Z_{(i+1)v})$$

$$= J(Z_1, \ldots, Z_v) + t_p J(Z_{v+1}, \ldots, Z_{2v}) + \cdots .$$

Just as in the proof of Proposition 35, H has no non-trivial zero in k_p, it has degree $p_1 + p_2 + p_3 = p$, and has pv variables. But

$$pv \geqq pp_1^\mu \geqq p(p^a)^\mu = p^{1+a\mu} = p_{\lambda(m, a)} ,$$

so the induction is complete.

The proof of (**) will use Vinogradov's famous theorem on the representation of numbers as sums of three primes. This theorem implies that there exists $b > 0$, such that the number of representations of the odd integer $N > b$ as $N = p_1 + p_2 + p_3$ with $p_1 \leqq p_2 \leqq p_3$, and all p_i prime exceeds $N^2/(\log N)^4$. Now clearly, the number of representations of N as $2n_1 + n_2$ for positive integers n_i is at most N, and the number of representations of N as $n_1 + n_2 + n_3$ for positive integers n_i with $n_1 \leqq N^\alpha$ is at most $N^{(1+\alpha)}$. Therefore, the number of representations of the odd integer $N > b$ as $p_1 + p_2 + p_3$ for prime p_i and $p_1^\alpha > N$ exceeds

$$N^2/(\log N)^4 - N - N^{1+\alpha} \geqq \frac{\text{polynomial in N of degree 2}}{(\log N)^4} .$$

This expression tends to ∞ as $N \to \infty$; therefore, there exists $c > 0$ such that if $N > c$, the number of representations as above is strictly positive. It follows that every prime number $> c$ is 1-representable. The general case follows by induction. Q.E.D.

Now to the counter-examples of Terjanian and Schanuel. Let us call an n-tuple of elements of $\mathbf{Z}_p$ —the p-adic integers—a *primitive* n-tuple [Gg], if one of the elements is a unit. Terjanian found a quartic form, f, in 3 variables over $\mathbf{Z}_2$ having the property that on primitive 3-tuples $\langle x_1, x_2, x_3 \rangle$

$$f(x_1, x_2, x_3) \equiv 1 \pmod 4 .$$

Granting the existence of this form for the moment, we may now use tricks similar to the above to construct the required counter-example. Namely if X, Y, Z represent distinct 3-tuples, then

$$g(X, Y, Z) = f(X) + f(Y) + f(Z)$$

is a quartic form in 9-variables, and obviously for any primitive 9-tuple $\langle x_1, x_2, x_3, y_1, y_2, y_3, z_1, z_2, z_3 \rangle$, we have

$$g(x, y, z) \equiv 1, 2, 3 \pmod 4 .$$

But then we claim that the quartic form in 18 (> 16) variables

$$(*) \qquad h(X, Y, Z, T, U, V) = g(X, Y, Z) + 2^2 g(T, U, V)$$

has no primitive zero mod 16; so it is the required counter-example. To see this, let $h(x, y, z, t, u, v) \equiv 0 \bmod 16$. Certainly then, $g(x, y, z) \equiv 0 \pmod 4$; hence, by the above $\langle x,y,z \rangle$ is not primitive. Now write $\langle x,y,z \rangle = 2\langle x', y', z', \rangle$ (as we may), then as g is quartic,

$$(**) \qquad g(x,y,z) = 16g(x', y', z') .$$

It follows from (*) and (**) that $g(t, u, v) \equiv 0 \pmod 4$ so that $\langle x,y,z,t,u,v \rangle$ cannot be primitive.

There remains the construction of the basic form f. This is given by

$$f(X_1, X_2, X_3) = \sum_{i=1}^{3} X_i^4 - \sum_{i,j=1}^{3} X_i^2 X_j^2 - \sum_{i,j,k=1}^{3} X_i^2 X_j X_k .$$

One finds that $f(1,0,0) = f(1,1,0) = 1$, $f(1,1,1) \equiv 1 \pmod 4$, and that $\partial f/\partial X_i \equiv 0 \pmod 2$ for $i = 1,2,3$. (Use the fact that every x in $\mathbf{Z}_2$ is congruent to its own square (mod (2).) If $\langle x_1, x_2, x_3 \rangle$ is primitive, write $x_j = y_j + 2x_j'$ where $y_j = 0, 1$ then compute $f(x) \pmod 4$ by Taylor's expansion. One finds that $f(x) \equiv 1 \pmod 4$, as required.

Schanuel's counter-examples are similar. He constructs a form f in two variables of degree $p(p-1)$ such that if $\langle x_1, x_2 \rangle$ is a primitive pair, then $f(x_1, x_2) \equiv 1 \pmod{p^2}$. Now take $p^2 - 1$ pairs of variables

$$Y_j = \langle X_{1j}, X_{2j} \rangle, \qquad j = 1, 2, \ldots, p^2 - 1$$

and make the form $g(Y_1, \ldots, Y_{p^2-1}) = \sum_{j=1}^{p^2-1} f(Y_j)$. Just as before, we find that

$$g(y_1, \ldots, y_{p^2-1}) \not\equiv 0 \pmod{p^2}$$

if $\langle y_1, \ldots, y_{p^2-1} \rangle$ is primitive. The actual counter-example form, h, is constructed from g by repeating the Terjanian trick:

$$h = g_0 + p^2 g_2 + p^4 g_4 + \cdots + p^{d-2} g_{d-2}, \qquad d = p(p-1)$$

where each g_j is g but with a new set of variables. Exactly the same argument as before shows that the form h of degree d in $p(p+1)(p-1)^2 > d^2$

variables has no primitive zero (mod p^2); so it is the required counter-example.

The form f in this case is $\phi(X^{p-1}, Y^{p-1})$, where

$$\phi(X,Y) = X^p + Y^p - \frac{1}{2}(X^{p-1}Y + XY^{p-1}) \; .$$

If $\langle x, y \rangle$ is a primitive pair then either both x, y are units or (say) x is a unit while $y = pz$. In the first case $x^{p-1} \equiv y^{p-1} \equiv 1 \pmod p$, while in the second case $y^{p-1} \equiv 0 \pmod{p^2}$. However, the basic property of the form ϕ is that if one of its variables is $\equiv 1$ mod p and the other is either congruent to 1 mod p or to zero mod p^2, then $\phi(x, y) \equiv 1 \pmod{p^2}$ (*). Since a primitive pair always has $p-1$st powers satisfying these conditions, the form f has the necessary property.

(*) To see this, use a case by case discussion. We may assume $x \equiv 1 + pz$, then either (a) $y = 1 + pt$ or (b) $y = p^2u$. If case (a), then $x^p + y^p \equiv 2 \pmod{p^2}$, $x^{p-1}y \equiv 1 + (u-t)p \pmod{p^2}$, while $xy^{p-1} \equiv 1 + (t-u)p \pmod{p^2}$. Thus, $\phi(x, y) \equiv 2 - ½(2) = 1 \pmod{p^2}$. If case (b), it follows immediately that $\phi(x, y) \equiv x^p \equiv 1 \pmod{p^2}$.

CHAPTER V

LOCAL CLASS FIELD THEORY

§1. Local fields and their extensions—a resumé

Definition 7. A *local field* is a complete, discretely valued field with finite residue class field.

One knows that local fields fall into two types:

(a) In characteristic zero, they are finite extensions of $\mathbf{Q}_p$—the rational p-adic field (which, in turn, is the fraction field of the Witt vectors on $\mathbf{Z}/p\mathbf{Z}$);

(b) In characteristic $p > 0$, they are formal power series fields in one variable with coefficients in a finite field.

Our standard notations concerning these fields will be: ord_k = exponential valuation on the local field k, it has value group $\mathbf{Z}$; $\tilde{k}$ will be the residue field; U will denote the *group of units* in $k = \{x \in k \mid \text{ord}_k(x) = 0\}$; $\mathcal{O}$ will denote the *ring of integers* in $k = \{x \in k \mid \text{ord}_k(x) \geqq 0\}$; $\mathfrak{p}$ will denote the maximal ideal of $\mathcal{O}$ (consisting of those $x \in \mathcal{O}$ for which $\text{ord}_k(x) > 0$).

One then has that

$$\mathfrak{p}^r = \{x \in k \mid \text{ord}_k(x) \geqq r\} = (\pi^r)$$

where π is a prime element of k, i.e., $\text{ord}_k(\pi) = 1$. The ideals (π^r), $r = 0, 1, 2, \ldots$ constitute *all* the non-zero ideals of $\mathcal{O}$, and

$$\mathcal{O}/\mathfrak{p} \xrightarrow{\sim} \tilde{k}; \quad \mathcal{O}/\mathfrak{p} \xrightarrow{\sim} \mathfrak{p}^r/\mathfrak{p}^{r+1}$$

the latter isomorphism being induced by $\xi \mapsto \pi^r\xi$. We also have the filtration

$$k^* \supseteq U \supseteq U_1 \supseteq \cdots \supseteq U_n \supseteq \cdots$$

where $U_n = 1 + \mathfrak{p}^n$ for $n \geqq 1$. In the additive filtration

$$k^+ \supseteq \mathcal{O} \supseteq \mathfrak{p} \supseteq \mathfrak{p}^2 \supseteq \cdots$$

the factor groups are $k^+/\mathcal{O}$, $\tilde{k}$, $\tilde{k}, \ldots$, while in the multiplicative filtration the factor groups are $\mathbf{Z}$, $\tilde{k}^*$, $\tilde{k}$, $\tilde{k}, \ldots$. The group $\mathcal{O}$ is compact and open in k^+, and U is compact and open in k^*.

If K/k is a finite extension of k, then there is a unique extension of the valuation ord_k to one on K and in this valuation K is a local field. If $\mathcal{O}_K$, $\mathfrak{p}_K$, U_K, etc. denote the various constructs for K, we have:

$$\mathfrak{p}_K \cap k = \mathfrak{p}_k, \quad \text{so } \tilde{K} \supseteq \tilde{k} \ .$$

Let $f_{K/k}$ be the degree of the residue class extension $[\tilde{K}: \tilde{k}]$. Let $e_{K/k}$ be defined by the equation

$$\mathrm{ord}_K(\pi_k) = e_{K/k}\, \mathrm{ord}_K(\pi_K) \ ,$$

then $e_{K/k}$ is the *ramification index* of K/k. It is not hard to see that $\mathcal{O}_K$ is a free $\mathcal{O}_k$-module with basis $\xi_i \pi_K{}^j$, $i = 1, 2, \ldots, f_{K/k}$, $j = 0, 1, \ldots, e_{K/k} - 1$, where the ξ_i are elements of $\mathcal{O}_K$ whose residue classes in $\tilde{K}$ are a basis for $\tilde{K}$ over $\tilde{k}$. Thus,

$$[K : k] = e_{K/k} f_{K/k} \ .$$

If $e_{K/k} = 1$, we call K/k an *unramified extension* (we always assume K/k is a separable extension); if $f_{K/k}$ is 1 we call K/k a *totally ramified extension.*

The main result used to study unramified extensions of a local field is Hensel's Lemma. We shall just state it, a proof may be found in [Gg].

PROPOSITION 36 (Hensel's Lemma). *Let* k *be a local field, let* $f \in \mathcal{O}[X]$, *and assume at least one coefficient of* f *is a unit of* k. *Let* $\tilde{a}$ *be a simple root of* $\tilde{f}$ *(so that* $\tilde{f} = (X - \tilde{a})\,\tilde{g}(X)$*), then there exists an* $a \in k$ *with residue class* $\tilde{a}$ *such that* a *is a simple root of* f.

Remark. A proof of Proposition 36 can be obtained from Lemma 8.

COROLLARY. *If* K/k *is a finite extension and if* k *is a local field, then there exists a unique subextension* L *of* K/k *which is unramified over* k *and has residue class field* $\tilde{K}$. *Each unramified extension* M *with* $K \supseteq M \supseteq k$ *is already contained in* L *and the Galois group of* L/k *is exactly that of* $\tilde{K}/\tilde{k}$. *The field extension* K/L *is purely ramified; hence, each finite extension* K/k *of the local field* k *has a composition series*

$$K \supseteq L \supseteq k$$

in which K/L *is purely ramified and* L/k *is unramified.*

If we pass to the direct limit in the corollary above over all finite normal extensions, then we obtain (as the direct limit of the fields L) the *maximal unramified extension,* k_T, *of k.* Its invariance group in G_k is denoted T. Another way of obtaining k_T and T is to note that as k is complete, there is a unique way of extending its valuation to k_s—the separable closure of k. Thus for all $\sigma \in G_k$, $\mathrm{ord}_{k_s}(\sigma a) = \mathrm{ord}_{k_s}(a)$ because $\mathrm{ord}_{k_s}(\sigma a)$ is another extension of $\mathrm{ord}_k(a)$ just as $\mathrm{ord}_{k_s}(a)$ is one. Hence, $\sigma\mathcal{O}_{k_s} = \mathcal{O}_{k_s}$, $\sigma\mathfrak{p}_{k_s} = \mathfrak{p}_{k_s}$; and it follows that G_k acts naturally on the residue field, $\tilde{k}_s$, of k_s. ($\tilde{k}_s$ is the separable (= algebraic) closure of $\tilde{k}$.) This gives us a *surjection* $G_k \to G_{\tilde{k}}$ whose kernel is the group T. The fixed field of T is k_T, and the residue field, $\tilde{k}_T$, of k_T, is precisely $(\tilde{k})_s$. Thus the factor group $G_k/T = G_{\tilde{k}}$ is $\hat{Z} = \operatorname{proj\,lim}_n Z/n!\,Z$, and an extension L of k is unramified if and only if $L \subseteq k_T$.

We say an extension K/k is *tamely ramified* iff $\mathrm{ch}\,\tilde{k}$ does not divide $e_{K/k}$. If $\sigma \in G_k$ and $a \in k_s^*$, then $\mathrm{ord}_{k_s}(a) = \mathrm{ord}_{k_s}(\sigma a)$, so the symbol

$$\langle \sigma, a \rangle = (\sigma a)/a \mod \mathfrak{p}_{k_s}$$

makes sense and has values in $\tilde{k}_s^*$. For fixed $a \in k_s^*$, $\langle \sigma, a \rangle$ is a 1-cocycle of G_k with values in $\tilde{k}_s^*$, so as T acts trivially on $\tilde{k}_s^*$, we see that $<\sigma, a>$ (for *fixed* a) is a homomorphism from T to $\tilde{k}_s^*$. Thus,

$$<\sigma_1 \sigma_2, a> = <\sigma_1, a> <\sigma_2, a> \qquad \text{for all } \sigma_j \in T.$$

Moreover, one finds trivially,

$$<\sigma, a_1 a_2> = <\sigma, a_1><\sigma, a_2> \qquad \text{for all } a_j \in k_s^*.$$

It follows that we have a pairing $T \times k_s^* \to \tilde{k}_s^*$. Let V be the kernel on the left in this pairing. That is, $V = \{\sigma \in T \mid <\sigma, a> = 1 \text{ for all } a \in k_s^*\}$. Equivalently,

$$V = \{\sigma \in T \mid (\forall a \in k_s^*)(\frac{\sigma a}{a} - 1 \in \mathfrak{p}_{k_s})\} = \{\sigma \in T \mid \mathrm{ord}_{k_s}(\sigma a - a) > \mathrm{ord}_{k_s}(a)$$
$$\text{for all } a \in k_s^*\}.$$

The group V is closed in T; hence it is closed in G_k, its fixed field will be denoted k_V. What is the kernel on the right? Observe that if $\mathrm{ord}_{k_s}(a) = \mathrm{ord}_{k_s}(b)$, there is a unit, say u, of k_s^* with $a = ub$. Thus, $\sigma a = \sigma u \sigma b$, or $\frac{\sigma a}{a} = \frac{\sigma u}{u} \frac{\sigma b}{b}$. But as $\sigma \in T$, $\sigma u \equiv u \bmod \mathfrak{p}$, i.e., $(\widetilde{\frac{\sigma u}{u}}) = 1$. We deduce that $\mathrm{ord}_{k_s}(a) = \mathrm{ord}_{k_s}(b)$ implies $<\sigma, a>$ equals $<\sigma, b>$ for all σ in T. Therefore the right hand kernel evidently contains U_{k_s}. But even more lies in the kernel. For if the b above lies in k, then $\sigma b = b$; hence $<\sigma, a> = 1$ in this case as well. From this discussion we obtain the non-degenerate pairing $<\sigma, a>$ between T/V and $\mathbf{Q}/\mathbf{Z} = \mathrm{ord}_{k_s}(k_s^*)/\mathrm{ord}_k(k^*)$ into the group $\tilde{k}_s^* =$ all n^{th} roots of 1 for n prime to $p = \mathrm{ch}\,\tilde{k}$. Observe that the identifications $\mathrm{ord}_{k_s}(k_s^*) \xrightarrow{\sim} \mathbf{Q}$ and $\mathrm{ord}_k(k^*) \xrightarrow{\sim} \mathbf{Z}$ depend upon our choice of prime elements and as such are non-canonical. Since the pairing is non-degenerate it gives (*via* duality) the isomorphism

$$T/V \xrightarrow{\sim} \prod_{q \neq p} Z_q \qquad \text{(non-canonical).}$$

Thus the Galois group of k_V/k_T is $\prod_{q \neq p} Z_q$, so that k_V is obtained from k_T by adjunction of all elements of the form $\pi^{1/m}$, where $(m, p) = 1$ and π is a prime element of k. It follows from this that k_V is a tamely ramified extension of k, obtained as the compositum of *Eisenstein extensions* of k_T (i.e., those extensions one gets by adjoining a root of the prime element π to k_T). We have now provèd the first few statements of

PROPOSITION 37. *The pairing* $T \times k_s^* \to \tilde{k}_s^*$ *given by the symbol* $<\sigma, a> = \widetilde{(\sigma a/a)}$ *establishes a non- canonical isomorphism between* $G(k_V/k_T)$ *and* $\Pi_{q \neq p} Z_q$. *Here,* k_V *is the fixed field of* V, *where* V *is the kernel on the left in the pairing* $<\sigma, a>$. *The field* k_V *is obtained from* k_T *by Eisenstein extensions, that is, adjunctions of* m^{th} *roots of the prime element* π *of* k *with* $(m, p) = 1$. *Finally,* k_V *is the maximal tamely ramified extension of* k*; that is, the group* V *is a profinite* p*-group.*

Proof: We need to prove only the last statement. Let K/k be a finite normal extension, let $L = K \cap k_T$ be the maximal unramified extension of K/k, and let $M = K \cap k_V$. We must show K/M is a p-extension. If $a \in K^*$, and if $\sigma \in V$ represents an element of $V/V \cap G_K$—the Galois group of K/M—then as $\sigma \in V$, we obtain $\text{ord}_{k_s}(\sigma a - a) > \text{ord}_{k_s}(a)$. Hence, $\text{ord}_{k_s}(\sigma - 1) a > \text{ord}_{k_s}(a)$. It follows that

$$\text{ord}_{k_s}(a) < \text{ord}_{k_s}(\sigma - 1) a < \text{ord}_{k_s}(\sigma - 1)^2 a < \cdots .$$

Now observe that $\text{ord}_{k_s}(\sigma - 1)^p a = \text{ord}_{k_s}(\sigma^p - 1) a$ (we may assume a is an integer of K, and then one uses the binomial theorem and reduction mod $\mathfrak{p}_K$). Hence, we obtain

$$\text{ord}_{k_s}(a) < \text{ord}_{k_s}(\sigma^p - 1)a < \text{ord}_{k_s}(\sigma^{p^2} - 1)a < \cdots .$$

Since σ^{p^n} mod $V \cap G_K$ ranges over a finite set, the ascending chain of ord's must stop and can do so only when one of the terms is infinite. For this term, $(\sigma^{p^r} - 1)a = 0$; and if we choose a to be a primitive element for

K/M (i.e., $K = M(a)$), we deduce $\sigma^{p^r} = 1$. Since σ was arbitrary, we are done. Q.E.D.

COROLLARY. *Every finite normal extension of a local field is solvable.*

Proof: Each finite normal extension has a tower: its maximal unramified extension, its maximal tamely ramified extension, itself. Let us denote this by

$$K \supseteq M \supseteq L \supset k.$$

The Galois groups $G(L/k)$, $G(M/L)$, $G(K/M)$ are respectively cyclic of order $f_{K/k}$, cyclic of order prime-to-p part of $e_{K/k}$, a p-group of order the p-primary part of $e_{K/k}$. Hence, all these Galois groups are solvable; therefore so is $G(K/k)$. Q.E.D.

As we have mentioned, $\mathcal{O}_k = \text{proj lim } \mathcal{O}_k/\pi^\nu \mathcal{O}_k$; so $\mathcal{O}_k$ is a profinite ring and as such it is compact. Since $\mathcal{O}_k$ is open in k, we see that k is locally compact. Let μ be a Haar measure on the additive group, k^+, of k–say μ normalized by $\mu(\mathcal{O}_k) = 1$. If $\alpha \neq 0$, $\alpha \in k$, then $x \mapsto \alpha x$ is an automorphism of k^+; hence it multiplies the Haar measure μ by some constant $\Delta(\alpha)$.

$$\mu(\alpha C) = \Delta(\alpha)\mu(C), \text{ for every measurable set } C.$$

Observe that $\mu(\alpha\beta C) = \mu(\alpha(\beta C)) = \Delta(\alpha)\mu(\beta C) = \Delta(\alpha)\Delta(\beta)\mu(C)$; while $\mu(\alpha\beta C) = \Delta(\alpha\beta)\mu(C)$. Thus $\Delta(\alpha)\Delta(\beta) = \Delta(\alpha\beta)$. If $\tilde{k}$ has q elements, the *normed absolute value of* α is defined by the equation

$$\|\alpha\|_k = \left(\frac{1}{q}\right)^{\text{ord}_k(\alpha)} = q^{-\text{ord}_k(\alpha)}.$$

PROPOSITION 38. *Let* k *be a local field with ring of integers* $\mathcal{O}_k$. *Then we have the equalities*

$$\|\alpha\|_k = \Delta(\alpha) \left(= \frac{1}{(\mathcal{O}_k : \alpha\mathcal{O}_k)}, \text{ if } \alpha \in \mathcal{O}_k \right)$$

for every $\alpha \in k^*$. *If* K/k *is a finite extension, and if* $\alpha \in k$, *then*

$$\|\alpha\|_K = \|\alpha\|_k^{[K:k]}$$

Proof: The set $\mathcal{O}_k$ is μ-measurable; we shall test the equalities above on the set $\mathcal{O}_k$. Given $\alpha \in k^*$, write $\alpha = \pi^r\beta$ with π a prime element of k and $\beta \in \mathcal{O}_k$. The multiplicativity of $\| \ \|_k$ and Δ shows that we may assume $\alpha \in \mathcal{O}_k$. This being said, write $\alpha = \pi^r\beta$ with β a unit—so that $r = \mathrm{ord}_k(\alpha) \geqq 0$. Then $\alpha\mathcal{O}_k = \pi^r\mathcal{O}_k$, and we know $\mathcal{O}_k/\pi^r\mathcal{O}_k$ is finite and of cardinal q^r. Hence,

$$\mu(\mathcal{O}_k) = q^r\mu(\alpha\mathcal{O}_k) = \|\alpha\|_k^{-1}\mu(\alpha\mathcal{O}_k) ,$$

as required. To prove the last statement, write $[K:k] = e_{K/k}f_{K/k}$. Now $\#(\tilde{K}) = q^{f_{K/k}}$, and $\mathrm{ord}_K(\alpha) = e_{K/k}\,\mathrm{ord}_k(\alpha)$. Thus,

$$\|\alpha\|_K = \left(\frac{1}{q^{f_{K/k}}}\right)^{\mathrm{ord}_K(\alpha)} = \left(\frac{1}{q}\right)^{e_{K/k}f_{K/k}\,\mathrm{ord}_k(\alpha)} = \|\alpha\|_k^{[K:k]} .$$

Q.E.D.

PROPOSITION 39. *Let* k *be a local field, and let* m *be a natural integer prime to the characteristic of* k. *Then for all* $\nu \geqq \mathrm{ord}_k m + 1$, *the mapping* $a \mapsto a^m$ *is an isomorphism of* U_ν *onto* $U_{\nu+\mathrm{ord}_k m}$. *That is,*

$$(U_\nu)^m = U_{\nu+\mathrm{ord}_k m} \quad \textit{for } \nu \geqq \mathrm{ord}_k m + 1.$$

Proof: $U_\nu = 1 + \mathfrak{p}^\nu = \{x \mid x = 1 + \pi^\nu y \text{ for some } y \in \mathcal{O}_k\}$. Clearly, $(U_\nu)^m \subseteq U_{\nu+\mathrm{ord}_k m}$. The kernel of $U_\nu \to (U_\nu)^m$ is the set of all m^{th} roots of 1 contained in U_ν. Let ξ be such a root of 1, then $\xi - 1 \in \mathfrak{p}^\nu$. Hence, $\xi^2 \equiv 1 \bmod \mathfrak{p}^\nu$, and in general, $\xi^r \equiv 1 \bmod \mathfrak{p}^\nu$. But, if $\xi \neq 1$,

$$\xi^{m-1} + \xi^{m-2} + \cdots + \xi + 1 = 0 ,$$

so we deduce $m \in \mathfrak{p}^{\nu}$. Thus $\mathrm{ord}_k(m) \geqq \nu > \mathrm{ord}_k m$, a contradiction. It follows that $U_{\nu} \to (U_{\nu})^m$ is a monomorphism. If $\alpha \in U_{\nu+\mathrm{ord}_k m}$, consider the equation $f(X) = X^m - \alpha = 0$. We apply Newton's Approximation Lemma (our Lemma 8) to this equation with $\alpha_0 = 1$. $f(\alpha_0)$ is then $1-\alpha$, while $f'(\alpha_0) = m$. Hence, $\mathrm{ord}_k(f(\alpha_0)/f'(\alpha_0)^2) = \nu + \mathrm{ord}_k(m) - 2\,\mathrm{ord}_k(m) > 0$. Lemma 8, yields an element $a \in \mathcal{O}_k$ with $f(a) = 0$, i.e., $a^m = \alpha$. Moreover, $\mathrm{ord}_k(a-\alpha_0) = \mathrm{ord}_k(a-1) \geqq \mathrm{ord}_k(f(\alpha_0)/f'(\alpha_0)) = \nu$; hence, $a \in U_{\nu}$, as required. Q.E.D.

COROLLARY. *For sufficiently large* ν, *we have* $U_{\nu}/U_{\nu}{}^m \simeq \mathcal{O}/m\mathcal{O}$; *hence* $(U_{\nu} : U_{\nu}^m) = (\mathcal{O} : m\mathcal{O}) = \|m\|_k^{-1}$. *In particular, "near one" is like "near zero" in* k.

Remark. A more precise statement of the last part of the corollary is the local isomorphism (topological) of k^* and k^+. This is effected by log and exp as is usual and the reader is invited to consider the domains of existence of their power series expansions.

§2. Cohomological Triviality, Tate and Nakayama Theorems, and Herbrand Quotients

We shall collect here many abstract statements of a cohomological flavor which were motivated by and have their main applications in Class Field Theory. They will be used repeatedly throughout the rest of the development.

The first topic is cohomological triviality which has already occurred in §3 of Chapter IV as well as earlier in the proof of the Šafarevič-Golod Theorem. Recall the following fact: If A is a G-module with $pA = (0)$, where G is a finite p-group (and p is a prime number, of course) then the following are equivalent (a) $A = (0)$, (b) $A^G = (0)$, (c) $A_G = H_0(G, A) = (0)$ (Rim's Lemma). Here is another lemma due to Rim.

LEMMA 11 (Rim). *Let* G *be a finite* p*-group, and let* A *be a* G*-module for which* $pA = (0)$. *If* $H_1(G, A) = (0)$, *then* A *is a free* $Z/pZ\,[G]$*-module; hence* $\tilde{H}^q(G, A) = (0)$ *for all* $q \in Z$.

Proof: Let $\Lambda = Z/pZ\,[G]$, let I be the augmentation ideal of Λ, and let $n =$ cardinal of $H_0(G, A)$. By Rim's first lemma, there is a free Λ-module F on n generators with

$$0 \to R \to F \to A \to 0$$

exact. Fact (c) above shows that $F/IF \xrightarrow{\simeq} A/IA$. Apply homology to the sequence above, we get

$$H_1(G, A) \to H_0(G, R) \to H_0(G, F) \to H_0(G, A) \to 0\,.$$

Since $H_1(G, A) = (0)$, we deduce that $R/IR = H_0(G, R) = (0)$. Now Rim's first lemma shows that $R = (0)$, so $F = A$. Q.E.D.

PROPOSITION 40. *Let* G *be a finite* p*-group, and let* A *be a* G*-module with* $pA = (0)$. *Then the following are equivalent.*

(a) $(\exists\, q \in Z)(\tilde{H}^q(G, A) = (0))$

(b) A *is cohomologically trivial,* i.e., $\tilde{H}^q(U, A) = (0)$ *for every* q *and every subgroup* U *of* G.

(c) $A = \pi_{*1 \to G}(B)$ *for some* B

(d) A *is a free* $Z/pZ\,[G]$*-module.*

Proof: (d) $\Longrightarrow$ (c) because $\pi_{*1 \to G}(Z/pZ) = \Lambda = Z/pZ[G]$.

(c) $\Longrightarrow$ (b) by Shapiro's Lemma

(b) $\Longrightarrow$ (a), this is trivial

(a) $\Longrightarrow$ (d). By dimension shifting, we can find a G-module B with $pB = (0)$ such that $\tilde{H}^n(G, A) = \tilde{H}^{n-q-2}(G, B)$. Then $H_1(G, B) = \tilde{H}^{-2}(G, B) = \tilde{H}^q(G, A) = (0)$ by (a). Lemma 11 shows that B is free; hence, $\tilde{H}^n(G, B) = (0)$ for all n. But then, $H_1(G, A) = \tilde{H}^{-2}(G, A) = \tilde{H}^{-q-4}(G, B) = (0)$, and Lemma 11 shows that A is free. Q.E.D.

PROPOSITION 41. *Let* G *be a finite* p-*group and let* A *be a* G-*module without* p-*torsion. Then the following are equivalent*

(1) $\tilde{H}^q(G, A) = (0)$ *for two consecutive dimensions*

(2) A *is cohomologically trivial*

(3) A/pA *is* Λ (= Z/pZ [G]) *-free.*

Proof: Write $0 \to A \xrightarrow{p} A \to A/pA \to 0$, and apply cohomology.

$$\cdots \to \tilde{H}^q(G, A) \xrightarrow{p} \tilde{H}^q(G,A) \to \tilde{H}^q(G, A/pA) \to \tilde{H}^{q+1}(G, A) \xrightarrow{p} \tilde{H}^{q+1}(G, A) \to \cdots$$

(3) $\Longrightarrow$ (2). By Proposition 40, and the above sequence (taken for any subgroup U of G), we find that multiplication by p is an isomorphism on $\tilde{H}^q(U, A)$ for every subgroup U of G and every $q \in Z$. Since G is a p-group, this implies (2).

(2) $\Longrightarrow$ (1) This is trivial.

(1) $\Longrightarrow$ (3). Let q, q + 1 be the consecutive dimensions for which $\tilde{H}^q(G, A)$ vanishes. By the cohomology sequence above, we deduce $\tilde{H}^q(G, A/pA) = (0)$, and Proposition 40 completes the proof. Q.E.D.

COROLLARY. *Let* A *be a* Z-*free module satisfying any (hence all) of conditions* (1), (2), (3) *of Proposition 41. If* B *is a torsion free* G-*module, then the* G-*module* $C = \mathrm{Hom}_Z(A, B)$ *is cohomologically trivial.*

Proof: Now C is torsion free, and we shall prove that C/pC is cohomologically trivial. Propositions 40 and 41 will then complete the proof. The exact sequence $0 \to B \xrightarrow[p]{} B \to B/pB \to 0$ together with the fact that A is free (over Z) yields the exact sequence

$$0 \to C \xrightarrow[p]{} C \to \mathrm{Hom}_Z(A, B/pB) \to 0 .$$

Thus $C/pC \xrightarrow{\sim} \mathrm{Hom}_Z(A, B/pB) \xrightarrow{\sim} \mathrm{Hom}_Z(A/pA, B/pB)$.

Condition (3) of Proposition 41 says that A/pA is Λ-free; hence, by Proposition 40, A/pA is $\pi_{*1 \to G}(D)$ for some D. It follows immediately that

C/pC is $\pi_{*1 \to G}(E)$ for some E; hence, by Proposition 40, C/pC is cohomologically trivial. Q.E.D.

PROPOSITION 42. *Let* G *be a finite group, and let* A *be a* **Z**-*free* G-*module. Let* G_p *denote a* p-*Sylow subgroup of* G. *The following two conditions are equivalent.*

(1) *For every* p, A *satisfies one of conditions* (1), (2), (3) (*hence, all of them*) *of Proposition 41 for the group* G_p.

(2) A *is* **Z**[G]-*projective.*

Proof: Clearly, (2) $\Longrightarrow$ (1); we shall prove (1) $\Longrightarrow$ (2). Write A as a quotient of a free **Z**[G]-module, say F, with kernel K,

$$0 \to K \to F \to A \to 0 \ .$$

Since A is **Z**-free, we obtain the exact sequence

$$(*) \qquad 0 \to C = \mathrm{Hom}_Z(A, K) \to \mathrm{Hom}_Z(A, F) \to \mathrm{Hom}_Z(A, A) \to 0 \ .$$

Now K is torsion free and A satisfies (1), so the Corollary above implies C is cohomologically trivial. Applying ordinary (non-reduced) cohomology to (*), we deduce

$$0 \to C^G \to \mathrm{Hom}_G(A, F) \to \mathrm{Hom}_G(A, A) \to H^1(G, C) = (0)$$

is exact. This shows that there exists a G-homomorphism $A \to F$ lifting the identity $A \to A$, that is, the exact sequence

$$0 \to K \to F \to A \to 0$$

splits. It follows that A is **Z**[G]-projective. Q.E.D.

Propositions 40 - 42 allow us to give a short proof of the "twin number" criterion mentioned in §3 of Chapter IV.

THEOREM 31 (Twin Number Criterion). *Let* G *be a finite group and let* A *be an arbitrary* G-*module. Then the following are equivalent:*

(1) *For every* p, *there exist two consecutive integers* i_p, $i_p + 1$ *such*

that $\tilde{H}^{i_p}(G_p, A) = \tilde{H}^{i_p+1}(G_p, A) = (0)$

(2) A *is cohomologically trivial.*

Proof: It will only be necessary to prove (1) $\Longrightarrow$ (2). Write A as the quotient of a free Z[G]-module F with kernel K. Since F is Z-free, K, being a submodule of a free Z-module, is also Z-free. Apply cohomology to the exact sequence

$$0 \to K \to F \to A \to 0 \ .$$

We get the exact sequence

$$0 \to \tilde{H}^{i_p}(G_p, A) \to \tilde{H}^{i_p+1}(G_p, K) \to 0 \to \tilde{H}^{i_p+1}(G_p, A) \to \tilde{H}^{i_p+2}(G_p, K) \to 0 \ ,$$

and hypothesis (1) shows that $\tilde{H}^q(G_p, K) = \tilde{H}^{q+1}(G_p, K) = (0)$ for suitable integers q, q + 1. According to Proposition 42, we may conclude that K is Z[G]-projective. From this, and the cohomology sequence of the exact sequence $0 \to K \to F \to A \to 0$, we deduce that A is cohomologically trivial.

Q.E.D.

Remark. If we replace "Z-free" by "divisible" in Proposition 42, and if we retain cohomological triviality for A in the statement of that Proposition, then we may deduce that A is Z[G]-injective. The proof is essentially dual to the one given.

There are several beautiful applications of these criteria for cohomological triviality to theorems useful in class field theory. We shall give two such applications.

THEOREM 32 (Nakayama). *Let* A *and* B *be* G-*modules for the finite group* G. *Assume that* A *is cohomologically trivial. In order that* $A \otimes B$ (*resp.* Hom(A, B), *resp.* Hom(B, A)) *be cohomologically trivial, it is necessary and sufficient that* $\mathrm{Tor}_1(A, B)$ (*resp.* $\mathrm{Ext}^1(A, B)$, *resp.* $\mathrm{Ext}^1(B, A)$) *be cohomologically trivial.*

Proof: From the proof of Theorem 31 we see that because A is cohomologically trivial, we can find a free Z[G]-module F and a projective

$\mathbf{Z}[G]$-module K so that $0 \to K \to F \to A \to 0$ is exact. Tensor this exact sequence (over $\mathbf{Z}$) with B. We obtain

$$0 \to \mathrm{Tor}_1(A,B) \to K \otimes B \to F \otimes B \to A \otimes B \to 0$$

is exact. Since K and F are respectively projective and free modules over G, their tensor products with B are direct summands of induced modules; hence $K \otimes B$ and $F \otimes B$ are cohomologically trivial. It follows immediately from an exact sequence argument that $A \otimes B$ is cohomologically trivial if and only if $\mathrm{Tor}_1(A,B)$ is. Similar remarks prove the statements involving Hom and Ext. Q.E.D.

COROLLARY (Nakayama). *If* A *is a cohomologically trivial* G-*module for the finite group* G *and if* B *is another* G-*module such that either* A *or* B *is torsion free (resp. either* A *is* $\mathbf{Z}$-*free or* B *is divisible), then* $A \otimes B$ *(resp.* $\mathrm{Hom}(A,B)$*) is cohomologically trivial.*

THEOREM 33 (Tate). *Let* G *be a finite group, let* $A \times B \to C$ *be a* G-*pairing of* G-*modules, and let* $a \in \tilde{H}^p(G,A)$ *be chosen. For each subgroup* U *of* G, *and for each* q *in* $\mathbf{Z}$, *the cup-product with* $\mathrm{res}_{G\to U}(a)$ *yields a homomorphism*

$$\theta_q(U\,;a): \tilde{H}^q(U,B) \longrightarrow \tilde{H}^{q+p}(U,C) \quad .$$

Assume for some q_0 *that (1)* $\theta_{q_0-1}(U\,;a)$ *is surjective for all* U, *(2)* $\theta_{q_0}(U\,;a)$ *is bijective for all* U, *(3)* $\theta_{q_0+1}(U;a)$ *is injective for all* U. *Then,* $\theta_q(U\,;a)$ *is an isomorphism for every* $q \in \mathbf{Z}$ *and every subgroup* U *of* G.

Proof: We have two exact sequences

$$0 \to A \to \pi_{*1\to G}(\pi^*_{1\to G}A) \to A'' \to 0$$

$$0 \to A' \to \mathbf{Z}[G] \otimes_{\mathbf{Z}} A \to A \to 0$$

which serve to dimension shift down, resptively up, in the cohomology sequences based on reduced cohomology. Our pairing $A \times B \to C$ yields pairings $A'' \times B \to C''$, $A' \times B \to C'$ where C'' is to C (resp. C' is to C) as A'' is to A (resp. A' is to A). Moreover, the theorem proved for the case of A', B, C' or A'', B, C'' yields the theorem for A, B, C by the isomorphisms of dimension shifting. Hence, we may and do assume that $p = 0$; that is, $a \in \tilde{H}^0(G, A)$.

In this case, a is represented by an element (again denoted a) of A^G, and the same element represents $\mathrm{res}_{G\to U}(a)$ for each subgroup U of G. The pairing $A \times B \to C$ yields a map $f\colon B \to C$ given by $f(b) = \langle a, b\rangle$, so that $\theta_q(U, a)$ is just the map

$$\tilde{H}^q(U, B) \to \tilde{H}^q(U, C)$$

induced by the map $f\colon B \to C$.

Write $0 \to B \xrightarrow[\varepsilon_B]{} \pi_{*1\to G}(\pi^*_{1\to G} B)$, and set $\overline{C} = C \amalg \pi_*(\pi^* B)$. (Observe the analogy to the mapping cylinder operation of topology.) Define $\overline{f}\colon B \to \overline{C}$ *via* the equation

$$\overline{f}(b) = f(b) \oplus \varepsilon_B(b) \quad ,$$

then $\overline{f}$ is *injective*. Since $\tilde{H}^q(C) = \tilde{H}^q(\overline{C})$ for every q, we may and do assume that $f\colon B \to C$ is injective.

After all these reductions, we may complete the proof using Theorem 31, as follows: We have the exact sequence

$$0 \to B \xrightarrow{f} C \to D \to 0 \,, \qquad D = \operatorname{coker} f \,,$$

and the cohomology sequence yields (with obvious notations)

$$\cdots \to \tilde{H}^{q_0-1}(B) \to \tilde{H}^{q_0-1}(C) \to \tilde{H}^{q_0-1}(D) \to \tilde{H}^{q_0}(B) \to \tilde{H}^{q_0}(C) \to \tilde{H}^{q_0}(D)$$
$$\to \tilde{H}^{q_0+1}(B) \to \tilde{H}^{q_0+1}(C) \to \cdots \,.$$

Our hypotheses show that for every subgroup U of G, $\tilde{H}^{q_0-1}(U, D) = \tilde{H}^{q_0}(U, D) = (0)$. If we let U run over the p-Sylow subgroups of G, we may apply Theorem 31 to conclude that $\tilde{H}^n(U, D)$ vanishes for every n and every U. Hence,

$$\tilde{H}^n(U, B) \longrightarrow \tilde{H}^n(U, C)$$

is an isomorphism for every n and every U, and this is the assertion of Tate's Theorem. Q.E.D.

Cup-products also give us a nice proof of the theorem (due to Artin and Tate) that cyclic groups have periodic cohomology—in fact the period is 2. This result is important for the subject of Herbrand quotients, which in turn is important for class field theory.

Suppose G is a finite group. The homology sequence for the exact sequence

$$0 \to I \to Z[G] \to Z \to 0$$

yields

$$0 \to H_1(G, Z) \to I/I^2 \to Z[G]/I \to Z/IZ \to 0 \ .$$

Since $Z[G]/I = Z$ and $Z/IZ = Z$, we deduce that $H_1(G, Z) \cong I/I^2$. Now the mapping $\sigma \longmapsto \sigma - 1$ is an isomorphism of $G^{ab} = G/[G, G]$ with I/I^2 as one may readily check. Hence,

$$\tilde{H}^{-2}(G, Z) = H_1(G, Z) \xrightarrow{\sim} G/[G, G] = G^{ab} \ .$$

We have shown in the discussion following Proposition 10, that the group $\tilde{H}^2(G, Z) = H^2(G, Z)$ is isomorphic to the character group of G^{ab} *via* the isomorphism

$$\hat{G}^{ab} = H^1(G, Q/Z) \xrightarrow[\sim]{\delta} H^2(G, Z) \ .$$

If G is a cyclic group of order n, then corresponding to a generator ζ of G we have two elements, one in $\tilde{H}^{-2}(G, Z)$, the other in $\tilde{H}^2(G, Z)$. They are given as follows: ζ_{-2}, the element of $\tilde{H}^{-2}(G, Z)$, is that element corresponding to the image of ζ in G^{ab}; and ζ_2 is that element of $\tilde{H}^2(G, Z)$ which is the image (under δ) of the unique character, χ, of G whose value at ζ is $1/n$ (mod Z). The cup product of ζ_{-2} with ζ_2 yields an element of $\tilde{H}^0(G, Z) = Z/nZ$. We may calculate this cup-product explicitly by the formula

$$\zeta_{-2} \cup \zeta_2 = \sum_{\sigma \in G} Z(\sigma, \zeta) = \sum_{\sigma \in G} \delta\chi(\sigma, \zeta)$$

where $Z(\alpha, \beta)$ is a two-cocycle representing ζ_2.*

THEOREM 34 (Artin-Tate). *The cohomology of a cyclic group is periodic of period 2. More explicitly, if* ζ *is a generator of* G, *then for any* G-*module* A, *the pairing* $\mathbf{Z} \times A \to A$ *(given by* $\langle n, a \rangle \mapsto n\,a$*) induces the mappings*

$$\tilde{H}^n(G, A) \to \tilde{H}^{n+2}(G, A), \quad \textit{cupping with } \zeta_2$$
$$\tilde{H}^{n+2}(G, A) \to \tilde{H}^n(G, A), \quad \textit{cupping with } \zeta_{-2}$$

which are mutually inverse isomorphisms.

Proof: Let $a \in \tilde{H}^n(G, A)$, then we need to show that $(a \cup \zeta_2) \cup \zeta_{-2} = a = (a \cup \zeta_{-2}) \cup \zeta_2$. By the associativity of cup products, we need only show $\zeta_2 \cup \zeta_{-2} = 1$, $\zeta_{-2} \cup \zeta_2 = 1$; and by the skew commutativity of cup-products, we need prove only one of these. Now by the above,

$$\zeta_{-2} \cup \zeta_2 = \sum_{\sigma \in G} \delta\chi(\sigma, \zeta) = \sum_{\sigma \in G} (\chi(\zeta) - \chi(\sigma\zeta) + \chi(\sigma))$$

$$= \sum_{\sigma \in G} \chi(\zeta) - \sum_{\sigma \in G} \chi(\sigma\zeta) + \sum_{\sigma \in G} \chi(\sigma) \qquad (\text{mod } n\mathbf{Z})$$

$$= n\chi(\zeta) \quad \text{mod } n\mathbf{Z}$$

$$= 1 \quad (\text{mod } n\mathbf{Z}). \qquad \text{Q.E.D.}$$

(The statement $\Sigma_{\sigma \in G}\, \chi(\sigma) = 0$ follows as χ is a non-trivial character.)

---

* The reader may dig a proof for this out of [CE, p. 243], or consult [SCL, pp. 184-186]. The proof in the latter reference is very valuable as it applies to give the "Nakayama map" for the reciprocity law.

We complete this *potpourri* of cohomological algebra with some material on Herbrand Quotients. Suppose

$$A^{\cdot} : A_{-1} \to A_0 \to A_1 \to \cdots \to A_n \to A_{n+1}$$

is a complex of *finite* abelian groups. We define the *Euler-Poincaré characteristic* of $A^{\cdot}$, denoted $\chi(A^{\cdot})$, by the formula

$$\chi(A^{\cdot}) = \prod_{i=0}^{n} |A_i|^{(-1)^i} , \qquad |A_i| = \#(A_i) .$$

(This definition is a very special case of a vastly more general procedure involving categories, Grothendieck groups, and Euler maps, see [LA] for some details.)

Now the complex $A^{\cdot}$ has cohomology groups $H^j(A^{\cdot})$ given by

$$H^j(A^{\cdot}) = \ker(A_j \to A_{j+1}) / \operatorname{Im}(A_{j-1} \to A_j)$$

for $j = 0, 1, \ldots, n$. The collection of abelian groups $H^j(A^{\cdot})$ for $j = 0, 1, \ldots, n$ forms a complex $H^{\cdot}(A^{\cdot})$ under the trivial coboundary mappings, and so it has an Euler-Poincaré characteristic:

$$\chi(H^{\cdot}(A^{\cdot})) = \prod_{i=0}^{n} |H^i(A^{\cdot})|^{(-1)^i} .$$

PROPOSITION 43. *If* $A^{\cdot}$ *is a complex such that* $\chi(A^{\cdot})$ *is defined, then* $\chi(H^{\cdot}(A^{\cdot}))$ *is defined and*

$$\chi(A^{\cdot}) = \chi(H^{\cdot}(A^{\cdot})) .$$

Moreover, if $B^{\cdot}$ *is a subcomplex of* $A^{\cdot}$, *then*

(*) $$\chi(B^{\cdot})\chi(A^{\cdot}/B^{\cdot}) = \chi(A^{\cdot}) ,$$

in the sense that if any two of the three are defined, so is the third and equality (*) holds.

Proof: This is an essentially trivial induction; we shall give a quick sketch of the method of proof and omit the simple details. Let $B_{n-1} =$ Im A_{n-2} in A_{n-1}. Then our complex $A^{\cdot}$ yields two complexes

$$A^{\cdot\prime} \qquad A_{-1} \to A_0 \longrightarrow \cdots \longrightarrow A_{n-2} \longrightarrow B_{n-1} \longrightarrow 0$$

$$A^{\cdot\prime\prime} \qquad 0 \longrightarrow B_{n-1} \longrightarrow A_{n-1} \longrightarrow A_n \longrightarrow A_{n+1} \quad .$$

By induction hypothesis, $\chi(A^{\cdot\prime}) = \chi(H^{\cdot}(A^{\cdot\prime}))$; $\chi(A^{\cdot\prime\prime}) = \chi(H^{\cdot}(A^{\cdot\prime\prime}))$. Upon computation of the various cohomology groups in question, one sees immediately that $\chi(H^{\cdot}(A^{\cdot})) = \chi(A^{\cdot})$. Equation (*) follows from the above and the long exact sequence of cohomology. Q.E.D.

COROLLARY. *If* $0 \to A_0 \to \cdots \to A_n \to 0$ *is exact, then* $\chi(A^{\cdot}) = 1$.

To apply the above, we let A be an arbitrary abelian group (not necessarily finite) and let θ, ϕ be endomorphisms of A such that $\theta\phi = \phi\theta = 0$. Then

$$A^{\cdot}_{\theta,\phi} : A \xrightarrow{\phi} A_0 \xrightarrow{\theta} A_1 \xrightarrow{\phi} A$$

is a two-step complex with $A_0 = A$, $A_1 = A$ as shown. Its cohomology is $H^0(A^{\cdot}_{\theta,\phi}) = \ker\theta/\operatorname{Im}\phi$; $H^1(A^{\cdot}_{\theta,\phi}) = \ker\phi/\operatorname{Im}\theta$. Let us denote $\ker\theta$ by A_θ, $\operatorname{Im}\theta$ by A^θ with similar notations for ϕ. *We shall assume the cohomology groups of* $A^{\cdot}_{\theta,\phi}$ *are finite,* and when this is the case we define the *Herbrand Quotient of* A *with respect to* θ, ϕ, denoted $Q_{\theta,\phi}(A)$, by

$$(39) \qquad Q_{\theta,\phi}(A) = \chi(H^{\cdot}(A^{\cdot}_{\theta,\phi})) = \frac{(A_\theta : A^\phi)}{(A_\phi : A^\theta)} \quad .$$

PROPOSITION 44 (Herbrand's Lemma). *If* A *is a finite group, then* $Q_{\theta,\phi}(A) = 1$. *If* $B \subseteq A$ *is stable under* θ *and* ϕ, *then* $Q_{\theta,\phi}(A) = Q_{\theta,\phi}(B)\,Q_{\theta,\phi}(A/B)$, *where this is understood to mean that if any two are defined so is the third and equality holds. Consequently, if* A/B *is finite,* $Q_{\theta,\phi}(A) = Q_{\theta,\phi}(B)$.

Proof: When A is finite, $\chi(A^{\cdot}_{\theta,\phi})$ is defined and is manifestly equal to 1. By Proposition 43, $Q_{\theta,\phi}(A) = 1$ in this case. If $B \subseteq A$ and is stable under θ, ϕ, then we have the exact sequence of complexes

$$0 \longrightarrow B^{\cdot}_{\theta,\phi} \longrightarrow A^{\cdot}_{\theta,\phi} \longrightarrow (A/B)^{\cdot}_{\theta,\phi} \longrightarrow 0 \ .$$

This exact sequence yields the "long" exact sequence of cohomology, and the corollary above shows that the last statement of Herbrand's Lemma holds. Q.E.D.

There is one case in which the Herbrand Quotient is extremely important. Let m be an integer, and let A be an abelian group. If ϕ is multiplication by m while θ is the zero endomorphism, we may form

$$\Phi_m(A) = Q_{0,m}(A) = \frac{(A : A^m)}{(A_m : (0))} \ .$$

We call $\Phi_m(A)$ the *trivial Herbrand Quotient* corresponding to m. If A is a finite abelian group, Herbrand's Lemma yields the well-known statement: $(A : A^m) = (A_m : (0))$.

PROPOSITION 45. *Let* k *be a local field, and let* m *be an integer prime to the characteristic of* k. *Then* k^{*m} *is of finite index in* k^*, *in fact this index is given by*

$$(k^* : k^{*m}) = \frac{m}{\|m\|_k} ((k^*)_m : 1) \ .$$

Proof: The statement of the proposition is merely that $\Phi_m(k^*) = m/\|m\|_k$. Consider the exact sequence

$$0 \longrightarrow U_k \longrightarrow k^* \longrightarrow Z \longrightarrow 0 \ .$$

According to Herbrand's Lemma, $\Phi_m(k^*) = \Phi_m(U_k)\Phi_m(Z)$. Since $\Phi_m(Z) = m$, we need only prove that $\Phi_m(U_k) = 1/\|m\|_k$. Now for any $\nu \geqq 1$, the exact sequence

$$0 \longrightarrow U_\nu \longrightarrow U_k \longrightarrow U_k/U_\nu \longrightarrow 0$$

and the fact that U_k/U_ν is finite show that $\Phi_m(U_k) = \Phi_m(U_\nu)$. However, Proposition 39 shows that $((U_\nu)_m : (1)) = 1$ for ν sufficiently large and $(U_\nu : U_\nu^m) = 1/\|m\|_k$ for the same ν. The proof is complete. Q.E.D.

Remark. k^{*m} is closed in k^*, and Proposition 45 shows that it is open as well. By contrast, if $p =$ characteristic of k, then k^{*p} is closed in k^* but it is not open. To see this, observe that $k = k_0((t))$, for a finite field k_0. Consider the sequence

$$u_n = 1 + t^p + t^{p^2} + \cdots + t^{p^n} + t^{p^n+1}$$

of elements of k^*. Clearly $u_n \notin k^{*p}$, yet $\lim_n u_n$ is $\Sigma_{j=0}^{\infty} t^{p^j}$ which is a pth power. It follows that the index $(k^* : k^{*p})$ is not finite. (However, the group k^*/k^{*p} is compact as one sees from the exact sequence

$$0 \longrightarrow U_k/U_k{}^p \longrightarrow k^*/k^{*p} \longrightarrow Z/pZ \longrightarrow 0\ .\ )$$

The connection of the general Herbrand Quotient with local fields is made *via* cyclic extensions as follows: Let K/k be a cyclic extension of the local field k, let G be the Galois group, and let A be a G-module. Define the integer $h(G, A)$ by the equation

$$h(G, A) = \frac{h^2(G, A)}{h^1(G, A)}, \text{ where } h^n(G, A) = |\tilde{H}^n(G, A)|\ .$$

Since G is cyclic, Theorem 34 shows that

$$(*) \qquad h(G, A) = \frac{h^0(G, A)}{h^{-1}(G, A)} = \frac{(A^G : \mathfrak{N}A)}{(A_{\mathfrak{N}} : IA)}\ .$$

If θ is the endomorphism of A given by multiplication by $\sigma - 1$, where σ is a generator of G, and if ϕ is the *norm endomorphism* $\mathfrak{N}$ (which is multiplication by $\Sigma_{j=0}^{n-1} \sigma^j$), then (*) shows that $h(G, A)$ is $Q_{\theta,\phi}(A)$. This being the case, we shall refer to $h(G, A)$ as originally defined (in terms of

h^2 and h^1) as the *Herbrand Quotient of* G, A. If G acts trivially on A, an immediate computation yields

$$h(G, A) = \Phi_n(A), \qquad n = |G|, \quad \text{A having trivial action.}$$

This explains the terminology "trivial Herbrand Quotient." The following proposition summarizes well known facts and our above remarks.

PROPOSITION 46. *Let* k *be a local field, and let* K/k *be a cyclic extension of degree* n *with Galois group* G. *Then*

(a) $$h(G, k^*) = \Phi_n(k^*) = \frac{n}{\|n\|_k}, \qquad (n, p) = 1$$

(b) $$h(G, K^*) = h^2(G, K^*) .$$

When G is a cyclic group of *prime* order, the general Herbrand quotient h(G, A) may be computed in terms of the trivial Herbrand quotient $\Phi_q(A)$, where $q = |G|$. This is a remarkable theorem due to Chevalley (and generalized by Tate) whose proof employed representation theory. The shortest proof (which we give below) is due to John Tate, and there is a third proof due to J.-P. Serre which gives more information than Tate's proof but is more involved. We need a lemma.

LEMMA 12. *Let* α, β *be commuting endomorphisms of* A *and assume that* $Q_{0,\alpha}(A)$ *and* $Q_{0,\beta}(A)$ *are defined. Then* $Q_{0,\alpha\beta}(A)$ *is defined, and*

$$Q_{0,\alpha\beta}(A) = Q_{0,\alpha}(A)\, Q_{0,\beta}(A) .$$

Proof: Since $\alpha\beta = \beta\alpha$, we have the inclusions $A \supseteq A^{\alpha} \supseteq A^{\alpha\beta}$, $A_{\alpha\beta} \supseteq A_{\alpha} \supseteq (0)$. Thus, one finds

$$Q_{0,\alpha\beta}(A) = \frac{(A : A^{\alpha\beta})}{(A_{\alpha\beta} : (0))} = Q_{0,\alpha}(A)\, \frac{(A^{\alpha} : A^{\alpha\beta})}{(A_{\alpha\beta} : A_{\alpha})} ,$$

provided $Q_{0,\alpha\beta}(A)$ exists. However, $\alpha(A_{\alpha\beta}) \cong A_{\alpha\beta}/A_{\alpha} \cong A_{\beta} \cap A^{\alpha}$;

hence $(A_{\alpha\beta} : A_\alpha) = (A_\beta \cap A^\alpha : (0)) \leqq (A_\beta : (0)) < \infty$. Moreover, the exact sequence

$$0 \longrightarrow A^\alpha \cap A_\beta \longrightarrow A^\alpha \xrightarrow{\beta} A^{\alpha\beta} \longrightarrow 0$$

shows that $Q_{0,\beta}(A^\alpha) = \dfrac{(A^\alpha : A^{\alpha\beta})}{(A_{\alpha\beta} : A_\alpha)}$; hence, all will be proved when we show that $Q_{0,\beta}(A^\alpha)$ is defined and equals $Q_{0,\beta}(A)$. But, the sequence

$$0 \longrightarrow A_\alpha \longrightarrow A \longrightarrow A^\alpha \longrightarrow 0$$

is exact, and as $\alpha\beta = \beta\alpha$, each of A, A^α, A_α is stable under $0, \beta$. By Herbrand's Lemma,

$$Q_{0,\beta}(A) = Q_{0,\beta}(A^\alpha)\, Q_{0,\beta}(A_\alpha)\ ;$$

$Q_{0,\beta}(A^\alpha)$ is thus defined and moreover $Q_{0,\beta}(A_\alpha) = 1$ because A_α is a finite group. Q.E.D.

THEOREM 35. *Let* G *be a cyclic group of order* q, q *a prime number, and let* A *be a* G*-module. Assume* $\Phi_q(A)$ *is defined, then* $\Phi_q(A^G)$ *and* $h(G, A)$ *are defined and we have the equality*

$$h(G, A)^{q-1} = \Phi_q(A^G)^q / \Phi_q(A)\ .$$

Proof: We first show that $\Phi_q(A^G)$ and $h(G, A)$ are defined. Let σ be a generator of G, and consider the exact sequence

$$(*) \qquad 0 \longrightarrow A^G = A_{1-\sigma} \longrightarrow A \xrightarrow{(1-\sigma)} A^{(1-\sigma)} \longrightarrow 0\ .$$

Now $A^{1-\sigma}$ is both subgroup and a factor group of A, so one sees trivially that $\Phi_q(A^{1-\sigma})$ is defined because $\Phi_q(A)$ is defined. By Proposition 44, we now obtain that $\Phi_q(A^G)$ is defined. Since

$$h(G, A^G) = \Phi_q(A^G)\ ,$$

exact sequence (*) and Proposition 44 (again) show that $h(G, A)$ is defined and equal to the product $h(G, A^G)\, h(G, A^{1-\sigma})$. Moreover, we have the equation

$$\Phi_q(A) = \Phi_q(A^G)\Phi_q(A^{1-\sigma}) \quad ;$$

it follows that

(**)
$$h(G, A)^{q-1} = \Phi_q(A^G)^{q-1} h(G, A^{1-\sigma})^{q-1} \quad ,$$

$$\Phi_q(A)^{q-1} = \Phi_q(A^G)^{q-1} \Phi_q(A^{1-\sigma})^{q-1} ,$$

$$= \frac{\Phi_q(A^G)^q \, \Phi_q(A^{1-\sigma})^q}{\Phi_q(A)}$$

Hence,

(***)
$$\Phi_q(A^G)^q / \Phi_q(A) = \frac{\Phi_q(A)^{q-1}}{\Phi_q(A^{1-\sigma})^q} = \frac{\Phi_q(A^G)^{q-1}}{\Phi_q(A^{1-\sigma})} \quad .$$

If we compare (**) and (***), we see that our theorem will hold iff

$$h(G, A^{1-\sigma})^{q-1} = \frac{1}{\Phi_q(A^{1-\sigma})}$$

Now the module $A^{1-\sigma}$ is annihilated by $1 + \sigma + \cdots + \sigma^{q-1}$; hence $A^{1-\sigma}$ may be regarded as a module over $\mathbf{Z}[X]/(1 + X + \cdots + X^{q-1})$. In other words, *we may treat* σ *as a primitive* q^{th} *root of one.* From the definition of Herbrand Quotients, we find

$$h(G, A^{1-\sigma}) = Q_{0,\, 1-\sigma}(A^{1-\sigma})^{-1} \quad ;$$

so we must prove that

$$\Phi_q(A^{1-\sigma}) = Q_{0,\, 1-\sigma}(A^{1-\sigma})^{q-1} \quad .$$

However, in the cyclotomic field of qth roots of unity we have the equation $q = (1-\zeta)^{q-1}\,\varepsilon$ for some unit ε, where ζ is a primitive q^{th} root of unity. Our remarks show that in the endomorphism ring of $A^{1-\sigma}$, we have $q = (1-\sigma)^{q-1}\,\varepsilon$ for some *automorphism* ε *of* $A^{1-\sigma}$. As ε is an automorphism of $A^{1-\sigma}$, Lemma 12 shows that $Q_{0,\,\varepsilon}(A^{1-\sigma}) = 1$. Another application of

Lemma 12 yields

$$\Phi_q(A^{1-\sigma}) = Q_{0,\,q}(A^{1-\sigma}) = Q_{0,\,1-\sigma}(A^{1-\sigma})^{q-1}\, Q_{0,\,\varepsilon}(A^{1-\sigma})$$
$$= Q_{0,\,1-\sigma}(A^{1-\sigma})^{q-1} \, . \qquad \text{Q.E.D.}$$

Here is a simple yet important corollary of Theorem 35.

COROLLARY. *Let* k *be a local field and let* K/k *be a cyclic extension of prime degree* q *(where* (q, ch k) = 1*). Then,* h(G, K*) = q, $H^2(K/k, K^*)$ *is cyclic of order* q, *and*

$$(k^* : \mathfrak{N}_{K/k} K^*) = q = [K : k] \, .$$

Proof: By Theorem 35,

$$h(G, K^*)^{q-1} = |H^2(K/k, K^*)| = \Phi_q(k^*)^q / \Phi_q(K^*)$$
$$= \frac{q^q}{\|q\|_k^q} \Big/ \frac{q}{\|q\|_K} = \frac{q^{q-1}\|q\|_K}{\|q\|_k^q} = q^{q-1} \, .$$

Hence, h(G, K*) = q; and by Hilbert Theorem 90 and Theorem 34, we get

$$(k^* : \mathfrak{N}_{K/k} K^*) = h(G, K^*) = [K : k] \, . \qquad \text{Q.E.D.}$$

§3. The Brauer Group of a Local Field

We will now put together Lang's Theorem (Theorem 27) with the material of §§1 and 2. This will yield the structure of Br(k) when k is a local field; and, as we shall see, this structure is all important for the reciprocity law and the local class field theory.

THEOREM 36. *Let* k *be a local field, then the inflation map*

$$\mathrm{Br}(k_T/k) \to \mathrm{Br}(k)$$

is an isomorphism. Equivalently, every central simple algebra over k *has an unramified splitting field.*

Proof: We have the inflation-restriction sequence

$$0 \longrightarrow \mathrm{Br}(k_T/k) \longrightarrow \mathrm{Br}(k) \longrightarrow \mathrm{Br}(k_T) \quad ;$$

however, by Theorem 27, the field k_T is C_1. (For the completion of k_T has as residue field the algebraic closure of $\tilde{k}$, and as the field k_T is unramified over k, it is discretely valued.) Now Corollary 1 of Proposition 33 shows that $\mathrm{Br}(k_T) = (0)$. Q.E.D.

The deceptively short proof of Theorem 36 (it is based on a great deal of prior work) belies its importance. Briefly put, it reduces the determination of Br(k) to the unramified case, and this will now be our main object of study. We need the following Lemma.

LEMMA 13. *Let* G *be a profinite group, let* U *be a filtered* G*-module, say* $U = U_0 \supseteq U_1 \supseteq U_2 \supseteq \cdots$, *and assume*

(1) $H^r(G, U_i/U_{i+1}) = (0)$ *for* $i \geq 0$, *fixed* r ;

(2) U *is complete in the Hausdorff topology defined by the* U_j.

Then $H^r(G,U) = (0)$.

Proof: Let f be a cocycle representing a given cohomology class $\alpha \in H^r(G,U)$. If we map $U \to U/U_1$, the cohomology class α goes to zero by hypothesis (1), so f becomes a coboundary in U/U_1. That is, there exists $g \in C^{r-1}(G,U)$ such that $f - \delta g \in Z^r(G,U_1)$. Let $f_1 = f - \delta g$, then the same argument applied to f_1 and the map $U_1 \to U_1/U_2$ shows that there exists $g_1 \in C^{r-1}(G,U_1)$ with $f_1 - \delta g_1$ in $Z^r(G,U_2)$. If $f_2 = f_1 - \delta g_1$ we may proceed as above. In this manner, we construct a sequence $\{g_n\}$, each g_n belonging to $C^{r-1}(G,U_n)$ such that $f_n = f - \Sigma_{j=0}^{n} \delta g_j$ belongs to $Z^r(G,U_n)$. If G_n is the sum $\Sigma_{j=0}^{n} g_j$, then hypothesis (2) shows that $G = \lim G_n = \Sigma_{j=0}^{\infty} g_j$ exists; and it shows moreover, that $\lim f_n = 0$. Since δ is continuous in this topology, we deduce $f - \delta G = \lim(f - \delta G_n) = \lim f_n = 0$. Q.E.D.

PROPOSITION 47. *Let* k *be a local field, let* K/k *be a finite normal extension, and let* U_K *be the units of* K. *Then* $H^1(K/k, U_K)$ *is cyclic of order* $e_{K/k}$. *Moreover, the following three conditions are equivalent.*

(1) K/k *is unramified;*

(2) $H^1(K/k, U_K) = (0)$;

(3) U_K *is cohomologically trivial.*

Proof: Introduce the notations: Z_K = value group of K, Z_k that of k. Both are isomorphic to $\mathbf{Z}$, but the mapping $k^* \hookrightarrow K^*$ induces a map $Z_k \to Z_K$ in such a way that $Z_K/Z_k \xrightarrow{\sim} \mathbf{Z}/e_{K/k}\mathbf{Z}$. This being said, write the exact sequence

$$0 \longrightarrow U_K \longrightarrow K^* \longrightarrow Z_K \longrightarrow 0$$

and apply cohomology. We get the exact sequence

$$0 \longrightarrow U_k \longrightarrow k^* \longrightarrow Z_K \longrightarrow H^1(K/k, U_K) \longrightarrow 0$$

Thus, $H^1(K/k,U_K) \xrightarrow{\sim} \mathbf{Z}/e_{K/k}\mathbf{Z}$, as required.

The equivalence of (1) and (2) now follows trivially. Clearly (3) $\Longrightarrow$ (2) $\Longleftrightarrow$ (1), so let us prove that (1) $\Rightarrow$ (3). If (1) holds, then as (2) follows, the theory of cohomological triviality shows that all we need prove is that $H^2(K/k, U_K)$ vanishes. But U_K is filtered by the descending chain $U_K \supseteq U_1 \supseteq U_2 \supseteq \cdots$ (see §1), and the factor groups are $\tilde{K}^*$ and $\tilde{K}^+$ (from the second stage on). Since K/k is unramified, $G(K/k) \xrightarrow{\sim} G(\tilde{K}/\tilde{k})$; so that

$$H^2(K/k, U_K/U_1) \xrightarrow{\sim} H^2(\tilde{K}/\tilde{k}, \tilde{K}^*) = Br(\tilde{K}/\tilde{k})$$

$$H^2(K/k, U_j/U_{j+1}) \xrightarrow{\sim} H^2(\tilde{K}/\tilde{k}, \tilde{K}^+) .$$

All of these groups vanish—the Brauer groups because $\tilde{k}$ is finite; hence is C_1. Now Lemma 13 completes our proof. Q.E.D.

COROLLARY. *Every unit of* k *is a norm from every unramified extension of* k; *in fact, it is the norm of a unit in these extensions.*

Proof: $\tilde{H}^0(K/k, U_K) = \tilde{H}^2(K/k, U_K) = H^2(K/k, U_K) = (0)$ by our proposition. Hence, $U_k = \mathfrak{N}_{K/k} U_K$. Q.E.D.

We can now determine the Brauer Group.

THEOREM 37 (Hasse). *Let* k *be a local field, then there exists a canonical isomorphism*

$$\mathrm{inv}_k : \mathrm{Br}(k) \xrightarrow{\sim} \mathbf{Q}/\mathbf{Z}$$

called the invariant map (or Hasse invariant). The mapping inv_k *satisfies the following commutative diagram*

$$\begin{array}{ccc} \mathrm{Br}(k) & \xrightarrow{\mathrm{inv}_k} & \mathbf{Q}/\mathbf{Z} \\ \Big\downarrow \mathrm{res} & & \Big\downarrow [K:k] \\ \mathrm{Br}(K) & \xrightarrow{\mathrm{inv}_K} & \mathbf{Q}/\mathbf{Z} \end{array}$$

where the map on the right hand side is multiplication by the degree. Consequently, $\mathrm{Br}(K/k)$ *is canonically isomorphic to* $\mathbf{Z}/n\mathbf{Z}$ *and has a canonical generator called the fundamental class; denoted* $\alpha_{K/k}$.

Remark. The property expressed by the diagram above is frequently quoted by the phrase: "restriction multiplies invariant by degree."

Proof: According to Theorem 36, all we need prove for the first statement is a canonical isomorphism $\mathrm{Br}(k_T/k) \to \mathbf{Q}/\mathbf{Z}$. Now the sequence

$$0 \longrightarrow U_{k_T} \longrightarrow k_T^* \longrightarrow \mathbf{Z}_{k_T} = \mathbf{Z}_k \longrightarrow 0$$

is exact, and the cohomology sequence yields

$$0 \longrightarrow H^2(k_T/k, U_{k_T}) \longrightarrow \mathrm{Br}(k_T/k) \longrightarrow H^2(k_T/k, \mathbf{Z}) \longrightarrow 0.$$

But, $H^2(k_T/k, U_{k_T}) = \underset{(K/k \text{ unramified})}{\mathrm{dir\ lim}} H^2(K/k, U_K) = (0)$; hence

$$\mathrm{Br}(k_T/k) \xrightarrow[\sim]{\mathrm{ord}} H^2(k_T/k, \mathbf{Z}).$$

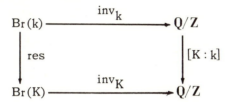

(Here, ord means "take the ordinal of a cochain".) However, as usual, there is the coboundary isomorphism

$$\delta : H^1(k_T/k, Q/Z) \longrightarrow H^2(k_T/k, Z) \ .$$

If q is the cardinal of $\tilde{k}$, then there is a canonical generator for $G(k_T/k)$- the Frobenius substitution F –given by $F(\alpha) \equiv \alpha^q \bmod \mathfrak{p}$. That is, $G(k_T/k)$ is canonically isomorphic to $\hat{Z}$, so the mapping $\chi \mapsto \chi(F)$ is a canonical isomorphism of $H^1(k_T/k, Q/Z)$ with Q/Z. If we put all these together we obtain inv_k, *viz*:

$$\underbrace{Br(k) \xrightarrow[\text{inf}^{-1}]{\sim} Br(k_T/k) \xrightarrow[\delta^{-1} \circ \text{ord}]{\sim} H^1(k_T/k, Q/Z) \xrightarrow{\sim} Q/Z}_{\text{inv}_k}$$

To prove that restriction multiplies invariant by degree, we first consider the cases of unramified and totally ramified extensions.

(a) If K/k is unramified, consider the diagram

$$\begin{array}{ccccccccc}
 & & & & \text{inv}_k & & & & \\
Br(k) & \xleftarrow{\text{inf}} & Br(k_T/k) & \xrightarrow{\text{ord}} & H^2(k_T/k, Z) & \xleftarrow{\delta} & H^1(k_T/k, Q/Z) & \longrightarrow & Q/Z \\
\downarrow \text{res} & & \downarrow \text{res} & & \downarrow \text{res} & & \downarrow \text{res} & & \downarrow [K:k] \\
Br(K) & \longleftarrow & Br(k_T/K) & \xrightarrow{\text{ord}} & H^2(k_T/K, Z) & \longleftarrow & H^1(k_T/K, Q/Z) & \longrightarrow & Q/Z) \\
 & & & & \text{inv}_K & & & &
\end{array}$$

Clearly, every square commutes except perhaps the last. But if F generates $G(k_T/k)$, and if $n = [K:k]$, then card $\tilde{K} = q^n$, so that F^n generates $G(k_T/K)$. Thus,

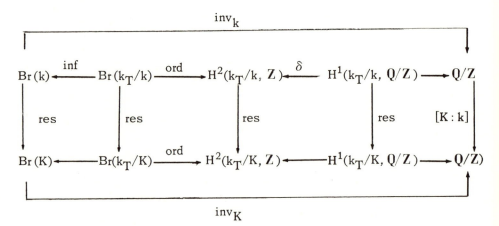

$$(\text{res}\,\chi)(F^n) = \chi(F^n) = n\chi(F) ,$$

and it follows that the last square commutes.

(b) If K/k is totally ramified, then k_T is linearly disjoint from K over k; hence one finds that K_T is the composite extension $k_T K$. We have the field diagram

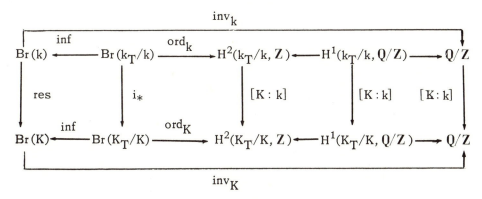

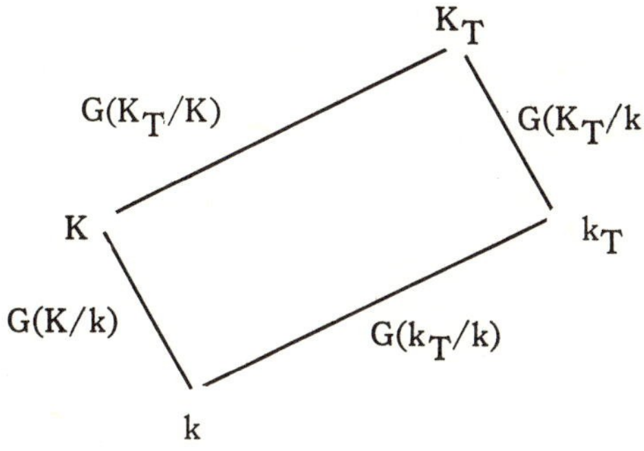

with "opposite sides of the parallelogram equal". This means that the injection $k_T^* \xrightarrow{i} K_T^*$ yields a homomorphism $\text{Br}(k_T/k) \xrightarrow[i_*]{} \text{Br}(K_T/K)$, and we have the diagram

$$\begin{array}{ccccccccc}
 & & & & \text{inv}_k & & & & \\
\text{Br}(k) & \xleftarrow{\text{inf}} & \text{Br}(k_T/k) & \xrightarrow{\text{ord}_k} & H^2(k_T/k, Z) & \longleftarrow & H^1(k_T/k, Q/Z) & \longrightarrow & Q/Z \\
\Big\downarrow \text{res} & & \Big\downarrow i_* & & \Big\downarrow [K:k] & & \Big\downarrow [K:k] & & \Big\downarrow [K:k] \\
\text{Br}(K) & \xleftarrow{\text{inf}} & \text{Br}(K_T/K) & \xrightarrow{\text{ord}_K} & H^2(K_T/K, Z) & \longleftarrow & H^1(K_T/K, Q/Z) & \longrightarrow & Q/Z \\
 & & & & \text{inv}_K & & & &
\end{array}$$

Clearly every square commutes save perhaps the middle one. However, K/k is totally ramified, so that $\text{ord}_K = [K:k]\,\text{ord}_k$ and the diagram commutes as claimed.

For the general case, we restrict in stages to the maximal unramified extension in K/k, then to K—the first step being case (a), the second, case (b). In the first step, case (a) shows inv is multiplied by $f_{K/k}$, and in the second, case (b) yields a further multiplication by $e_{K/k}$. Since $e_{K/k}f_{K/k} = [K:k]$, the general case is established.

Finally, the commutative diagram

$$\begin{array}{ccccccc}
0 \longrightarrow & Br(K/k) & \xrightarrow{\text{inf}} & Br(k) & \xrightarrow{\text{res}} & Br(K) \\
& \Big\downarrow{\scriptstyle inv_{K/k}} & & \Big\downarrow{\scriptstyle inv_k} & & \Big\downarrow{\scriptstyle inv_K} \\
0 \longrightarrow & \frac{1}{n}Z/Z & \longrightarrow & Q/Z & \xrightarrow{n} & Q/Z
\end{array}$$

where $n = [K:k]$ yields the last two statements of the theorem. Observe that $inv_{K/k}(\alpha_{K/k}) \equiv \frac{1}{n} \bmod Z$. Q.E.D.

If K/k is a normal extension with Galois group $G(K/k)$, and if $\sigma \in G_k = G(k_s/k)$, then for a subextension L of K/k, the mapping $\tau \mapsto \sigma^{-1}\tau\sigma$ is an isomorphism of $G(\sigma K/\sigma L)$ onto $G(K/L)$. Moreover, the mapping $a \mapsto \sigma a$ taking $K \to \sigma K$ is compatible with the conjugation; so the pair induces a map called *translation by* σ

$$\sigma_* : H^r(K/L, K^*) \longrightarrow H^r(\sigma K/\sigma L, \sigma K^*)$$

for each $r \geqq 0$. Observe that if $\sigma|L = 1$, we have a map of $H^r(K/L, K^*)$ to itself (since $\sigma K = K$) which is 1 in dimension zero; so, by universality, $\sigma_* = 1$ for all $r \geqq 0$ in this case.

Here are the functorial properties satisfied by inv_k and $\alpha_{K/k}$.

Proposition 48. *Let* K/k *be a normal extension of the local field* k, *and let* $\alpha_{K/k}$ *denote its fundamental class. Let* L/k *be a subextension of* K/k, *then the following properties hold:*

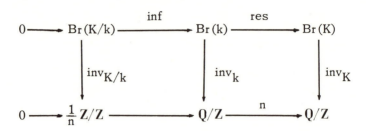

(1) $\mathrm{res}_{k\to L}\, \alpha_{K/k} = \alpha_{K/L}$

(2) $\mathrm{infl}\, \alpha_{L/k} = [K:L]\, \alpha_{K/k}$

(3) $\mathrm{inv}_K \circ \mathrm{res}_{k\to K} = [K:k]\, \mathrm{inv}_k$

(4) $\mathrm{inv}_k = \mathrm{inv}_K \circ \mathrm{tr}_{K\to k}$

(5) $\mathrm{tr}_{L\to k}\, \alpha_{K/L} = [L:k]\, \alpha_{K/k}$

(6) $\mathrm{inv}_{\sigma L} \circ \sigma_* = \mathrm{inv}_L$

(7) $\alpha_{\sigma K/\sigma L} = \sigma_*(\alpha_{K/L})$.

Proof: We have shown (3) in the proof of Theorem 37; all will follow from this assertion.

(1) We know that $\mathrm{inv}_k(\alpha_{K/k}) \equiv \dfrac{1}{[K:k]} \bmod \mathbf{Z}$,

and $$\mathrm{inv}_L(\alpha_{K/L}) \equiv \frac{1}{[K:L]} \bmod \mathbf{Z}$$

uniquely determine the α's. Thus,

$$\mathrm{inv}_L(\mathrm{res}_{k\to L}\, \alpha_{K/k}) = [L:k]\, \mathrm{inv}_k(\alpha_{K/k}) \equiv \frac{[L:k]}{[K:k]} \bmod \mathbf{Z}$$

$$\equiv \frac{1}{[K:L]} \bmod \mathbf{Z} = \mathrm{inv}_L(\alpha_{K/L}) \;\; ;$$

and property (1) follows.

(2) The diagram

$$\begin{array}{ccccccc}
0 & \longrightarrow & \mathrm{Br}(L/k) & \xrightarrow{\ \mathrm{inf}\ } & \mathrm{Br}(k) & \xrightarrow{\ \mathrm{res}\ } & \mathrm{Br}(L) \\
 & & \downarrow{\scriptstyle \mathrm{inf}} & & \downarrow{\scriptstyle \mathrm{Id}} & & \downarrow{\scriptstyle \mathrm{res}} \\
0 & \longrightarrow & \mathrm{Br}(K/k) & \xrightarrow{\ \mathrm{inf}\ } & \mathrm{Br}(k) & \xrightarrow{\ \mathrm{res}\ } & \mathrm{Br}(K)
\end{array}$$

commutes; and this shows that $\mathrm{inf}(\alpha_{L/k})$ and $[K:L]\alpha_{K/k}$ have the same invariant. This proves (2)

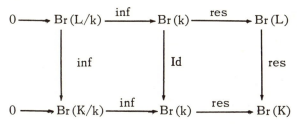

(3) Done

(4) Follows from (3) and tr $\circ$ res = index

(5) Follows from (1) and tr $\circ$ res = index

(6) We know that $\mathrm{res}_{k \to L}$ maps $\mathrm{Br}(k)$ *onto* $\mathrm{Br}(L)$ by Theorem 37. Moreover, because σ_* is a map arising from a compatible pair, we see easily that σ_* commutes with the appropriate restriction. This being said, let $x \in \mathrm{Br}(L)$, and let $y \in \mathrm{Br}(k)$ be chosen with $\mathrm{res}_{k \to L}(y) = x$. Then

$$\mathrm{inv}_{\sigma L}(\sigma_* x) = \mathrm{inv}_{\sigma L}(\sigma_* \mathrm{res}_{k \to L}(y))$$

$$= \mathrm{inv}_{\sigma L}(\mathrm{res}_{k \to \sigma L}(\sigma_* y)) = \mathrm{inv}_{\sigma L}\, \mathrm{res}_{k \to \sigma L}(y)$$

$$= [\sigma L : k]\, \mathrm{inv}_k(y) = [L : k]\, \mathrm{inv}_k(y) = \mathrm{inv}_L(\mathrm{res}_{k \to L}(y))$$

$$= \mathrm{inv}_L(x), \text{ as contended.}$$

(7) This follows immediately from (6) and the characterization of $\alpha_{K/k}$ in terms of inv_k. Q.E.D.

We can now combine the abstract theorem of Tate (our Theorem 33) with Hasse's result (Theorem 37) on the Brauer group to obtain what might be called the "main theorem of local class field theory".

THEOREM 38 (Tate). *Let* k *be a local field, let* K/k *be a finite normal extension, and let* $\theta : K^* \times \mathbf{Z} \to K^*$ *be the usual pairing* $\langle a, n\rangle \mapsto a^n$. *Then cup-product with the canonical class* $\alpha_{K/k}$ *yields the* ISOMORPHISMS

$$\theta^n : \tilde{H}^n(K/k, \mathbf{Z}) \longrightarrow \tilde{H}^{n+2}(K/k, K^*)$$

FOR EVERY $n \in \mathbf{Z}$. *These isomorphisms commute with restriction and transfer, and are multiplied by the index under inflation.*

Proof: Since restriction preserves canonical classes, the hypothesis of Theorem 33 dealing with "all subgroups U of G" is automatically satisfied. (Observe, as well, that θ is a $G(= G(K/k))$-pairing.) Now we must find the integer q_0 of the hypotheses of Theorem 33. Let $q_0 = 0$, then Hilbert Theorem 90 shows that $\tilde{H}^1(K/k, K^*)$ vanishes so that θ^{-1} is obviously surjective; while the result $\tilde{H}^1(K/k, Z) = (0)$ implies that θ^1 is injective. Therefore, all we need show is that

$$\theta^0 : \tilde{H}^0(K/k, Z) \longrightarrow Br(K/k)$$

is bijective. However, we know $\tilde{H}^0(K/k, Z)$ is cyclic of order $[K : k]$, and Theorem 37 says that the same is true of $Br(K/k)$. Moreover, the generator of $\tilde{H}^0(K/k, Z)$, namely 1, goes by cup-product with $\alpha_{K/k}$ to the element $\alpha_{K/k}$ of $Br(K/k)$ which is a generator of $Br(K/k)$ by Theorem 37 again. This immediately implies θ^0 is an isomorphism, and we are done.

The functorial properties of θ^n are immediate from those of $\alpha_{K/k}$ as expressed in Proposition 48 (1), (2), (5) together with minor checking in dimension zero. Q.E.D.

While Theorem 38 has many applications, its most crucial application is in the case $n = -2$. In this case, as we have remarked in §2, $\tilde{H}^{-2}(K/k, Z)$ is $G(K/k)^{ab}$ —the abelianized Galois group of K/k; while $\tilde{H}^0(K/k, K^*)$ is $k^*/\mathfrak{N}_{K/k}K^*$. The fundamental theorem yields the *isomorphism*

$$G(K/k)^{ab} \longrightarrow k^*/\mathfrak{N}_{K/k}K^*$$

via cup-product with the fundamental class $\alpha_{K/k}$. This isomorphism had been explicitly obtained by T. Nakayama [N] without the recognition that it was a cup-product or that it was one (the most important to be sure) of a series of such isomorphisms. In recognition of Nakayama's work, the mapping θ^{-2} is usually called the *Nakayama map.* The inverse isomorphism of θ^{-2}, that is, the isomorphism

$$k^*/\mathfrak{N}_{K/k}K^* \xrightarrow{\sim} G(K/k)^{ab}$$

is called the *reciprocity law* (or *reciprocity map*) of local class field theory. If $b \in k^*$, the composed mapping

$$k^* \longrightarrow k^*/\mathfrak{N}_{K/k}K^* \longrightarrow G(K/k)^{ab}$$

is called the *norm-residue map,* and the image of b in $G(K/k)^{ab}$ is denoted (b, K/k). The symbol (b, K/k) is called the *Artin symbol* or *norm-residue symbol.* Observe that the reciprocity law implies that the norm-residue symbol (b, K/k) is 1 if and only if b is a norm from K.

If K/k is a finite extension of the local field k, then the *norm-index of* K *over* k is the integer $(k^* : \mathfrak{N}_{K/k}K^*)$. Observe that the reciprocity law shows that this is *always finite.* For if L/k is finite and K/k is the smallest Galois extension containing L/k, then as $\mathfrak{N}_{K/k}(K^*) = \mathfrak{N}_{L/k}(\mathfrak{N}_{K/L}K^*) \subseteq \mathfrak{N}_{L/k}L^*$, we see that

$$(k^* : \mathfrak{N}_{L/k}L^*) \leqq (k^* : \mathfrak{N}_{K/k}K^*) = |G(K/k)^{ab}| < \infty .$$

We shall show in the next section that the subgroups $\mathfrak{N}_{K/k}K^*$ for finite K/k are always open subgroups of k^*.

The reciprocity law and invariant map are intimately connected, as we see from

PROPOSITION 49. *Let* K/k *be a normal extension and let* χ *be a character of* $G(K/k)^{ab}$ *(that is,* $\chi \in H^1(K/k, \mathbf{Q}/\mathbf{Z})$*). If* $a \in k^*$, *then*

$$\chi((a, K/k)) = \mathrm{inv}_k(a \cup \delta\chi) .$$

Hence, a knowledge of the invariant map implies a knowledge of the reciprocity law, and conversely.

Proof: First observe that for elements ξ of $\tilde{H}^{-2}(K/k, \mathbf{Z})$ and elements χ of $H^1(K/k, \mathbf{Q}/\mathbf{Z})$, the cup-product $\xi \cup \chi$ is given by $\chi(\xi)$. This being said, we know by definition that

$$(a, K/k) \cup \alpha_{K/k} = a ,$$

so that

$$((a, K/k) \cup \alpha_{K/k}) \cup \delta\chi = a \cup \delta\chi .$$

Hence,

$$a \cup \delta\chi = (-1)^{-4}\alpha_{K/k} \cup ((a, K/k) \cup \delta\chi)$$

$$= \alpha_{K/k} \cup ((a, K/k) \cup \delta\chi) \ .$$

Now $\delta(\xi \cup \chi) = \xi \cup \delta\chi \pm \delta\xi \cup \chi$ for any ξ, χ; so we deduce

$$a \cup \delta\chi = \alpha_{K/k} \cup \delta((a, K/k) \cup \chi)$$

$$= \alpha_{K/k} \cup \delta(\chi(a, K/k))$$

by our opening remarks. However, δ maps $\tilde{H}^{-1}(K/k, \mathbf{Q}/\mathbf{Z})$ to $\tilde{H}^{0}(K/k, \mathbf{Z})$ and is induced by the norm map, $\mathfrak{N}_{K/k}$. Since $G(K/k)$ acts trivially on $\mathbf{Q}/\mathbf{Z}$ and $\mathbf{Z}$, we find

$$\delta(\chi(a, K/k)) = [K:k]\chi(a, K/k) \in \mathbf{Z} \ .$$

It follows that

$$a \cup \delta\chi = \alpha_{K/k} \cup [K:k]\chi(a, K/k)$$

$$= ([K:k]\chi(a, K/k))(\alpha_{K/k} \cup 1)$$

$$= [K:k]\chi(a, K/k)\alpha_{K/k} \in H^2(K/k, K^*) \ .$$

And now we obtain

$$\mathrm{inv}_k(a \cup \delta\chi) = [K:k]\chi(a, K/k)\ \mathrm{inv}_k(\alpha_{K/k}) = \chi(a, K/k) \ .$$

Q.E.D.

Propositions 48 and 49 yield the following functorial behavior for the reciprocity law.

PROPOSITION 50. *Let* K/k *be a normal extension of the local field* k *and let* L *be an intermediate field. Then the diagrams* (on the following page) *commute, where the vertical maps are the reciprocity law, and the horizontal maps are as indicated.*

$$(1)\qquad \begin{array}{ccc} L^* & \xrightarrow{\mathfrak{N}_{L/k}} & k^* \\ \downarrow & & \downarrow \\ G(K/L)^{ab} & \xrightarrow[\text{can}]{} & G(K/k)^{ab} \end{array}$$

$$(2)\qquad \begin{array}{ccc} k^* & \xhookrightarrow{\text{incl.}} & L^* \\ \downarrow & & \downarrow \\ G(K/k)^{ab} & \xrightarrow{\text{tr}} & G(K/L)^{ab} \end{array}$$

Norm-transfer Theorem

$$(3)\qquad \begin{array}{ccc} L^* & \xrightarrow{\sigma} & \sigma L^* \\ \downarrow & & \downarrow \\ G(K/L)^{ab} & \xrightarrow[\tau \to \sigma\tau\sigma^{-1}]{} & G(\sigma K/\sigma L)^{ab} \end{array}$$

$\sigma \in G(k_s/k)$.

$$(4)\qquad \begin{array}{ccc} k^* & \xrightarrow{1} & k^* \\ \downarrow & & \downarrow \\ G(K/k)^{ab} & \xrightarrow[\text{proj}]{} & G(L/k)^{ab} \end{array}$$

if K *and* L *are normal*

Remarks 1. Diagram (3) may be written in the form:

$$\sigma(a, K/L)\sigma^{-1} = (\sigma a, \sigma K/\sigma L) \ .$$

2. According to diagram (4) we may pass to the limit over all finite normal extensions, and obtain the *reciprocity law mapping*

$$\theta_k = (-, k) : k^* \longrightarrow G_k^{ab} \ (= \text{Galois group of the maximal abelian extension of the local field } k.)$$

We write (a, k) or $\theta_k(a)$ for the image of the element a of k^* in the group G_k^{ab}. Upon passing diagrams (1) and (2) of Proposition 50 to the limit, we obtain the *Norm-Transfer Theorem: The diagram*

$$\begin{array}{ccc} L^* & \underset{\text{incl.}}{\overset{\mathfrak{N}}{\rightleftarrows}} & k^* \\ \theta_L \downarrow & & \downarrow \theta_k \\ G_L^{ab} & \underset{\text{tr}}{\overset{\text{can}}{\rightleftarrows}} & G_k^{ab} \end{array}$$

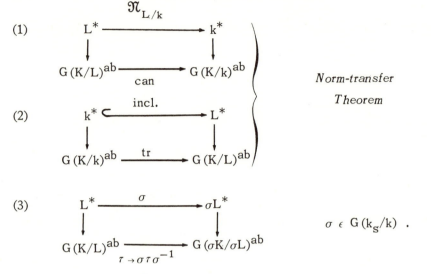

commutes for all algebraic extensions L *of the local field* k. Observe as well that diagram (3) in the form of Remark 1 becomes the equation:

$$\sigma(a, L)\sigma^{-1} = (\sigma a, \sigma L)$$

3. If we pass Proposition 49 to the limit, we obtain the characterization of θ_k *via* inv_k given by:

$$\chi(\theta_k(a)) = \mathrm{inv}_k(a \cup \delta\chi)$$

for every $\chi \in H^1(k, \mathbf{Q}/\mathbf{Z})$ *and every* $a \in k^*$.

The problem of explicitly calculating the symbol (a, K/k) for a given layer K/k and a given element a of k^* is very important. When K/k is unramified, the solution is classical and rather easy to obtain. (Proposition 51 below.) However, when K/k is purely ramified a local solution (i.e., not using global class field theory) is rather difficult and was first obtained by Dwork [Dk] and later, in more generality, by Lubin and Tate [LT]. The latter paper shows that methods of formal Lie groups are useful in local class field theory, and that ramification–theoretic phenomena can be explained in their terms. Here is the solution in the unramified case.

PROPOSITION 51. *Let* K/k *be a finite unramified extension of the local field* k. *Let* F *be the Frobenius automorphism–the canonical generator of* G(K/k) *–and let* $a \in k^*$ *be chosen. Then* (a, K/k) *is* $F^{\mathrm{ord}_k(a)}$.

Proof: Let χ be any character of the abelian group G(K/k). We must show that $\chi((a, K/k)) = \mathrm{ord}_k(a)\chi(F)$. However, Proposition 49 shows us that $\chi((a, K/k))$ is given by $\mathrm{inv}_k(a \cup \delta\chi)$. Now $a \cup \delta\chi$ is the cohomology class in Br(K/k) represented by the cocycle

$$f(\sigma, \tau) = a^{\delta\chi(\sigma, \tau)} \quad ;$$

and by definition of inv_k (cf. Theorem 37), we must take the ord_k of this cocycle, "remove" the δ, and apply the result to F. We obtain

$$\mathrm{ord}_k(a \cup \delta\chi) = \delta\chi(\sigma, \tau) \cdot \mathrm{ord}_k(a) = \delta(\mathrm{ord}_k\, a \cdot \chi) \quad ;$$

so that

$$\mathrm{inv}_k(a \cup \delta\chi) = \mathrm{ord}_k(a)\chi(F) ,$$

as contended. Q.E.D.

Observe that the reciprocity law and Proposition 51 show anew that every unit of k is a norm from any unramified extension.

§4. The Existence Theorem (Part I: Formal Existence Theorem, and beginning of the proof.)

The main theme of class field theory is the classification of the over fields of k by "arithmetic" objects within k itself. A means for doing this for local fields is *via* the reciprocity law—and this is its importance. We shall study this problem thoroughly in this section and the next, and shall accomplish the goal of classification as far as our methods will allow. The limitation arises naturally from:

PROPOSITION 52. *Let* K/k *be an extension of the local field* k, *and let* K^{ab} *be the maximal abelian subextension of* K/k. *Then* $\mathfrak{N}_{K/k}K^* = \mathfrak{N}_{K^{ab}/k}(K^{ab})^*$.

Remark. Proposition 52 shows that using the reciprocity law, one cannot hope to classify anything other than abelian extensions.

Proof: Since $K \supseteq K^{ab}$, we deduce $\mathfrak{N}_{K/k}K^*$ is contained in $\mathfrak{N}_{K^{ab}/k}(K^{ab})^*$. Let $\xi \in \mathfrak{N}_{K^{ab}/k}(K^{ab})^*$, and let L be a normal extension of k which contains K. Let $G = G(L/k)$, $H = G(L/K)$, then we know that $H[G, G] = G(L/K^{ab})$ and that $G(K^{ab}/k) = G/H[G, G]$. Since ξ is a norm from K^{ab}, the reciprocity law says that $(\xi, K^{ab}/k) = 1$ in $G(K^{ab}/k) = G/H[G, G]$. Consider $(\xi, L/k) \in G/[G, G]$, then projection onto $G/H[G, G]$ corresponds to forming $(\xi, K^{ab}/k)$ by diagram (4) of Proposition 50. It follows that $(\xi, L/k)$ lies in $\ker(G/[G, G] \to G/H[G, G])$, that is, that

$$(\xi, L/k) \in H/[H,H] = G(L/K)^{ab} \ .$$

Now the diagram

$$\begin{array}{ccc} K^* & \xrightarrow{\quad\mathfrak{N}\quad} & k^* \\ \big\downarrow & & \big\downarrow \\ G(L/K)^{ab} = H/[H,H] & \xrightarrow{\quad\text{can}\quad} & G/[G,G] = G(L/k)^{ab} \end{array}$$

commutes, and since the reciprocity law is surjective, we deduce that there is some $\eta \in K^*$ with

$$(\xi, L/k) = (\mathfrak{N}_{K/k}\eta, L/k) \ .$$

But the kernel of the norm-residue mapping is the norm group $\mathfrak{N}_{L/k} L^*$; hence, there exists $\zeta \in L^*$ with

$$\xi = \mathfrak{N}_{K/k}(\eta)\mathfrak{N}_{L/k}(\zeta) \ .$$

But then, $\xi = \mathfrak{N}_{K/k}(\eta \cdot \mathfrak{N}_{L/K}(\zeta)) \in \mathfrak{N}_{K/k} K^*$. Q.E.D.

If we know a norm group, $\mathfrak{N}_{K/k} K^*$, from an *abelian* extension K/k, we know the Galois group of that extension; hence, we know the extension. Therefore the classification of abelian over fields of k will follow from the classification of those subgroups of k^* which are norm groups. This is the subject of the existence theorem.

Recall that a map of topological spaces $f: X \to Y$ is called *proper* if the inverse image of any compact set is compact. One should think of proper maps as expressing the fact that "X is compact relative to Y"; in fact, if Y is a point, to say X is compact is the same as saying f is proper.

PROPOSITION 53. *Let K/k be a finite extension of the local field k. Then $\mathfrak{N}_{K/k}$ is a continuous, proper map from $K^* \to k^*$, and the group $\mathfrak{N}_{K/k} K^*$ is an open subgroup of finite index in k^*.*

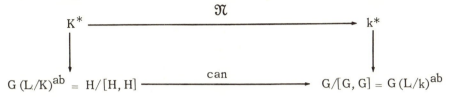

Proof: If $\alpha \in K^*$, then $\mathfrak{N}_{K/k}(\alpha)$ is the product of the conjugates of α (over k) with certain multiplicities. Since the Galois group $G_k = G(k_s/k)$ acts continuously, it follows that $\mathfrak{N}_{K/k}$ is a continuous mapping. We have the diagram

$$\begin{array}{ccccccccc}
0 & \longrightarrow & U_K & \longrightarrow & K^* & \xrightarrow{\text{ord}_K} & Z & \longrightarrow & 0 \\
 & & \Big\downarrow \mathfrak{N}_{K/k} & & \Big\downarrow \mathfrak{N}_{K/k} & & \Big\downarrow \mathfrak{N}_{K/k} & & \\
0 & \longrightarrow & U_k & \longrightarrow & k^* & \xrightarrow{\text{ord}_k} & Z & \longrightarrow & 0
\end{array}$$

which clearly commutes. If V is a compact subset of k^*, then the image of V in Z is compact; hence it is finite. Moreover, $V \cap U_k = V'$ is a compact subset of U_k, and as U_K is compact it follows that $\mathfrak{N}_{K/k}^{-1}(V')$ is a compact subset of K^*. But V is a finite union of translates of V' (since $\text{ord}_k(V)$ is finite); so we deduce that $\mathfrak{N}_{K/k}^{-1}(V)$ is a finite union of translates of the compact set $\mathfrak{N}_{K/k}^{-1}(V')$, that is, $\mathfrak{N}_{K/k}^{-1}(V)$ is compact.

Since $\mathfrak{N}_{K/k}$ is a proper map by the above, and since proper maps of locally compact Hausdorff spaces are always closed,[*] it follows that $\mathfrak{N}_{K/k}K^*$ is closed in k^*. We know from the reciprocity law that $\mathfrak{N}_{K/k}K^*$ is of finite index in k^*; hence, $\mathfrak{N}_{K/k}K^*$ is open in k^*. Q.E.D.

PROPOSITION 54. *If* K *and* L *are finite abelian extensions of the local field* k, *then* $\mathfrak{N}_{LK/k}(LK)^* = \mathfrak{N}_{L/k}(L^*) \cap \mathfrak{N}_{K/k}(K^*)$ *and* $\mathfrak{N}_{(L\cap K)/k}(L \cap K)^* = \mathfrak{N}_{L/k}L^* \cdot \mathfrak{N}_{K/k}K^*$.

Proof: Let $\xi \in \mathfrak{N}_{LK/k}(LK)^*$, then by transitivity of the norm we find that $\xi \in \mathfrak{N}_{L/k}L^*$ and $\xi \in \mathfrak{N}_{K/k}K^*$. Conversely, if

$$\xi \in \mathfrak{N}_{L/k}L^* \cap \mathfrak{N}_{K/k}K^* \; ,$$

(*) The reader may easily prove this himself, or he may apply the properness argument of the proof "in reverse" to show that $\mathfrak{N}_{K/k}K^*$ is closed.

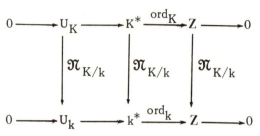

then the reciprocity law shows that $(\xi, L/k) = 1$ and $(\xi, K/k) = 1$. It follows that the element $(\xi, LK/k)$ when restricted to L and to K is the identity. But $G(L/k)$ and $G(K/k)$ generate $G(LK/k)$; hence, $(\xi, LK/k) = 1$ in $G(LK/k)$. The reciprocity law shows that $\xi \in \mathfrak{N}_{LK/k}(LK)^*$, as required.

For the second statement, the transitivity of the norm shows once again that

$$\mathfrak{N}_{L/k}L^* \cdot \mathfrak{N}_{K/k}K^* \subseteq \mathfrak{N}_{(L \cap K)/k}(L \cap K)^* \ .$$

If $\xi \in \mathfrak{N}_{L \cap K/k}(L \cap K)^*$, the reciprocity law shows that $(\xi, L \cap K/k) = 1$ in $G(L \cap K/k)$; hence, the elements $(\xi, L/k)$ and $(\xi, K/k)$ map onto 1 when restricted to the field $L \cap K$. However, the reciprocity law establishes an isomorphism $k^*/\mathfrak{N}_{K/k}K^* \xrightarrow{\sim} G(K/k)$ and the subgroup $G(K/L \cap K)$ corresponds to the subgroup $(\mathfrak{N}_{L/k}L^* \cdot \mathfrak{N}_{K/k}K^*)/\mathfrak{N}_{K/k}K^*$ by the Norm-transfer theorem. Now $(\xi, K/k)$ when restricted to $(L \cap K)$ is 1; so $(\xi, K/k) \in G(K/L \cap K)$, hence, $\xi \in \mathfrak{N}_{L/k}L^* \cdot \mathfrak{N}_{K/k}K^*$. Q.E.D.

Proposition 54 shows that the correspondence $K \rightsquigarrow \mathfrak{N}_{K/k}K^*$ is a 1-1 order inverting correspondence (lattice anti-isomorphism) between the abelian over fields of k and the set of norm subgroups of k. It also allows us to take the set of norm subgroups of k as a basis for open sets at the identity in k^*, and so to topologize anew the group k^*—the resultant topology is the *norm-topology* on k^* (or *class-topology*). Clearly the map $k^* \to k^*$ (the former with its usual topology, the latter with the norm topology) is continuous. Since $k^*/\mathfrak{N}_{K/k}K^* \xrightarrow{\sim} G(K/k)$ when K/k is abelian, we will complete k^* with respect to the norm topology to get

$$\hat{k}^* = \operatorname*{proj\,lim}_{K} k^*/\mathfrak{N}_{K/k}K^* \ ,$$

and we will call $\hat{k}^*$ the *norm completion of* k^*. Of course, the reciprocity law gives us an isomorphism

$$\theta_k : \hat{k}^* \longrightarrow G(k_s/k)^{ab} \ .$$

Now we can complete k^* with respect to the topology of open subgroups of finite index, so that there is an injection

$$k^* \longrightarrow \text{proj lim} \{k^*/U \mid U \text{ open}, (k^* : U) < \infty\} .$$

The existence theorem is essentially the statement that the two completions are the same. There is the canonical homomorphism $k^* \to \hat{k}^*$ whose kernel is the kernel of the norm-residue symbol $k^* \to G(k_s/k)^{ab}$ – that is, the group of *universal norms,* $\mathfrak{D}_k$,

$$\mathfrak{D}_k = \bigcap_K \mathfrak{N}_{K/k} K^* .$$

The norm topology is Hausdorff, and $k^* \to \hat{k}^*$ is injective if and only if the group of universal norms is trivial. This is a necessary condition for the validity of the existence theorem, and turns out to be the decisive step.

PROPOSITION 55. *Let* K/k *be a finite extension of the local field* k. *Let* $\mathfrak{D}_K$ *and* $\mathfrak{D}_k$ *be the groups of universal norms for* K, *respectively* k, *then* $\mathfrak{N}_{K/k}\mathfrak{D}_K = \mathfrak{D}_k$.

Proof: If $\xi \in \mathfrak{N}_{K/k}\mathfrak{D}_K$, then given L/k, we must prove $\xi \in \mathfrak{N}_{L/k}L^*$. Let $\Omega = KL$, then $\mathfrak{N}_{\Omega/K}\Omega^* \supseteq \mathfrak{D}_K$ and $\xi \in \mathfrak{N}_{K/k}(\mathfrak{N}_{\Omega/K}\Omega^*) \subseteq \mathfrak{N}_{\Omega/k}\Omega^* = \mathfrak{N}_{L/k}(\mathfrak{N}_{\Omega/L}\Omega^*)$. So $\xi \in \mathfrak{N}_{L/k}(L^*)$, as required.

Conversely, suppose $\xi \in \mathfrak{D}_k$. Given L, a finite extension of K, let

$$\Omega_\xi(L) = \mathfrak{N}_{L/K} L^* \cap \mathfrak{N}_{K/k}^{-1}(\xi) .$$

Since $\mathfrak{N}_{K/k}$ is a proper map, the sets $\Omega_\xi(L)$ are compact for all fields L over K. Moreover, because ξ is a universal norm, ξ is an element of $\mathfrak{N}_{L/k}(L^*)$ for each field L over K. If $\xi = \mathfrak{N}_{L/k}(\eta_L)$ for some $\eta_L \in L^*$, then clearly, $\mathfrak{N}_{L/K}(\eta_L) \in \Omega_\xi(L)$; so the sets $\Omega_\xi(L)$ are non-empty. Proposition 54 shows that the $\Omega_\xi(L)$ have the finite intersection property; hence by compactness there exists $\eta \in \bigcap_L \Omega_\xi(L)$. Thus, $\xi = \mathfrak{N}_{K/k}(\eta)$ and $\eta \in \mathfrak{D}_K$. Q.E.D.

To prove the triviality of $\mathfrak{D}_k$, we must make appeal to the arithmetic structure of k in a non-trivial way. This is the only point in the argument which is not essentially formal or readily generalized. The information we need is contained in the following Proposition, whose proof we will give in §5.

PROPOSITION 56. *Let* k *be a local field and let* K *be a finite extension of* k *containing the* p^{th} *roots of unity for some fixed prime* p. *If* $\xi \in K^*$, *and if* ξ *is a norm from every cyclic extension of* K *of degree* p, *then* $\xi \in K^{*p}$.

COROLLARY. *Let* k *be a local field, let* p *be a prime number, and let* $k_{(p)}$ *denote the field obtained from* k *by adjoining the* p^{th}*-roots of* 1 *to* k. *If* $L \supseteq k_{(p)}$ *is a field extension, then the map* $L^* \to L^*$ *via raising to the* p^{th} *power has compact kernel and its image contains* $\mathfrak{D}_L$.

Proof: The kernel is finite; hence compact. As for the image, Proposition 56 applies to the field L; so if $\xi \in \mathfrak{D}_L$ then $\xi \in L^{*p}$. Q.E.D.

PROPOSITION 57. *If* k *is local, the group of universal norms,* $\mathfrak{D}_k$, *is divisible. Hence,* $\mathfrak{D}_k = \bigcap_n k^{*n} = \{1\}$.

Proof: Choose any prime number p, all we need show is that $\mathfrak{D}_k^p = \mathfrak{D}_k$. If $\alpha \in \mathfrak{D}_k$, look at the set of all fields L finite over k such that $L \supseteq k_{(p)}$. Set

$$\Omega'_\alpha(L) = \{\beta \in k^* \mid \beta^p = \alpha \text{ and } \beta \in \mathfrak{N}_{L/k} L^*\} .$$

The sets $\Omega'_\alpha(L)$ are finite, hence compact. Since α is a universal norm, Proposition 55 shows that $\alpha \in \mathfrak{N}_{L/k}\mathfrak{D}_L$; and the Corollary to Proposition 56 shows that $\alpha \in \mathfrak{N}_{L/k}(L^{*p})$. Hence, $\alpha = \beta^p$ with $\beta \in \mathfrak{N}_{L/k}L^*$; and we have shown that $\Omega'_\alpha(L)$ is non-empty for each L. Once again Proposition 54 shows that the sets $\Omega'_\alpha(L)$ have the finite intersection property; by compactness there is some element $\beta \in \bigcap_L \Omega'_\alpha(L)$. Such an element β belongs to $\mathfrak{D}_k$ and $\beta^p = \alpha$; therefore $\mathfrak{D}_k$ is indeed divisible.

Since $k^* \supseteq \mathfrak{D}_k$, we deduce $k^{*n} \supseteq \mathfrak{D}_k$ for each n; hence $\bigcap_n k^{*n} \supseteq \mathfrak{D}_k$. However, an element of $\bigcap_n k^{*n}$ can only be a unit, and now Proposition 39 shows such an element must be 1. Q.E.D.

COROLLARY. *The norm-topology on* k^* *is Hausdorff, and the norm residue symbol yields a continuous injection of* k^* *into* $G(k_s/k)^{ab}$.

PROPOSITION 58 (Ramification Theorem). *Every subgroup of finite index in* k^* *which contains the group of units,* U_k, *is a norm group from an unramified extension. A necessary and sufficient condition that* K/k *be an unramified extension is that* $\mathfrak{N}_{K/k}K^* \supseteq U_k$.

Proof: We know that every unit of k is a norm from *every* unramified extension. Let I be a subgroup of k^* of finite index which contains U_k, say $(k^* : I) = n$. Then I corresponds to *the* subgroup $n\mathbf{Z}$ of $\mathbf{Z}$, i.e., I is uniquely determined by its index. Let K/k be the unique unramified extension of k of degree n, then $(k^* : \mathfrak{N}_{K/k}K^*) = [K:k] = n$. Moreover, our opening remarks show that $U_k \subseteq \mathfrak{N}_{K/k}K^*$; hence, $I = \mathfrak{N}_{K/k}K^*$. The second statement is now trivial. Q.E.D.

It is now a simple matter to prove the existence theorem.

THEOREM 39 (Existence Theorem). *A necessary and sufficient condition that a subgroup of* k^* *be a norm group is that it be open (hence closed) and of finite index. Consequently, there is a lattice anti-isomorphism between the set of closed subgroups of* k^* *of finite index and the set of abelian over-fields of* k. *The field corresponding to a given subgroup is called its class field, and the reciprocity law gives the structure of the Galois group of each abelian over field of* k. *The reciprocity law establishes an isomorphism of* k^* *completed in the topology of open subgroups of finite index with the Galois Group of the maximal abelian extension of* k.

Proof: We need to prove only that each open subgroup of finite index is a norm group. The other statements are merely a *resume of the* material of this section. Let I be an open subgroup of k^* of finite index, say $(k^* : I) = n$. Then $k^{*n} \subseteq I$; hence $\mathfrak{D}_k \subseteq I$.† Therefore, we deduce

$$\bigcap_K (\mathfrak{N}_{K/k} K^* \cap U_k) = \mathfrak{D}_k \cap U_k \subseteq \mathfrak{D}_k \subseteq I ;$$

since $\mathfrak{N}_{K/k} K^* \cap U_k$ is compact for each K, and since I is open, there is some K such that $\mathfrak{N}_{K/k} K^* \cap U_k \subseteq I$. By the modular law,

$$\mathfrak{N}_{K/k} K^* \cap U_k (\mathfrak{N}_{K/k} K^* \cap I) \subseteq I .$$

Since $\mathfrak{N}_{K/k} K^* \cap I$ has finite index, we see that $U_k \cdot (\mathfrak{N}_{K/k} K^* \cap I)$ is a subgroup of finite index in k^* containing U_k; hence it is a norm group by Proposition 58. Then, $\mathfrak{N}_{K/k} K^* \cap (U_k \cdot (\mathfrak{N}_{K/k} K^* \cap I))$ is also a norm group by Proposition 54, so that I contains a norm group. Now it is a triviality to verify that I is indeed a norm group. Q.E.D.

COROLLARY. *The diagram below is commutative, the left vertical arrow is an isomorphism, and the right vertical arrow is the canonical inclusion.*

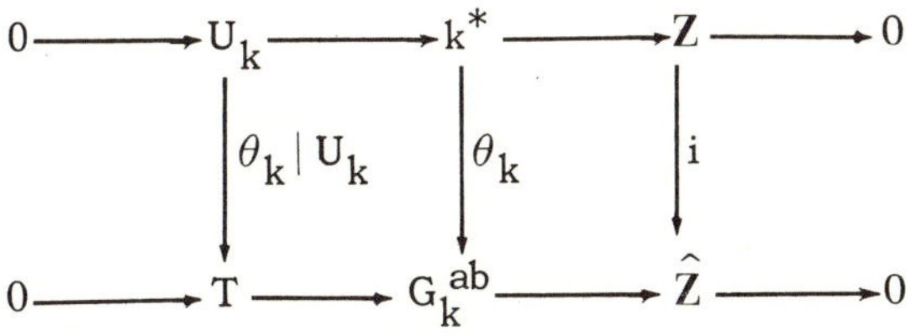

§5. The Existence Theorem (Part II. The Hilbert Symbols and completion of the proof.)

We now face the task of proving Proposition 56, and to do this we must introduce certain pairings which connect with the reciprocity law, the invariant map and the Brauer Group. In the next chapter we shall see that

† In the general abstract case, $\mathfrak{D}_k$ need not be zero for the proof to work, merely the divisibility of $\mathfrak{D}_k$ will do.

these pairings are cup-products and give rise to certain duality statements.

To begin with, let k be local, let $b \in k^*$, and let χ be a character of G_k (i.e., $\chi \in H^1(G_k, Q/Z) = H^1(G_k^{ab}, Q/Z)$). Then $\delta\chi$ belongs to $H^2(G_k, Z)$, and the natural pairing $k_s^* \times Z \to k_s^*$ *via* $\langle\lambda, n\rangle \mapsto \lambda^n$ yields a cup-product $b \cup \delta\chi \in H^2(G_k, k_s^*)$. Thus from b, χ we get an element of Br(k), say $\langle\chi, b\rangle$. Of course, the bilinearity of the cup-product gives us the equations

$$\langle\chi, b_1 b_2\rangle = \langle\chi, b_1\rangle + \langle\chi, b_2\rangle$$

$$\langle\chi_1 + \chi_2, b\rangle = \langle\chi_1, b\rangle + \langle\chi_2, b\rangle \ .$$

We can obtain an element of Q/Z from b and χ *via* $\{\chi, b\} = \mathrm{inv}_k\langle\chi, b\rangle$. In this form, Proposition 49 yields:

$$\{\chi, b\} = \mathrm{inv}_k(b \cup \delta\chi) = \chi(\theta_k(b)) \ ,$$

so that knowledge of the pairing $\{\chi, b\}$ for all characters χ is the same as determination of the reciprocity law. A pairing of the form $\{\chi, b\}$ (or one of several modifications of this form) is usually called a *local symbol,* and its connection with Artin's reciprocity law gives it its importance. We will know the reciprocity mapping for certain finite layers if we know the local symbol $\{\chi, b\}$ for certain characters.

The first class of characters arises from Kummer Theory, and the corresponding (modified) local symbol is usually called the Hilbert Symbol. Let n be an integer prime to the characteristic of k, and assume that k contains a primitive n^{th} root of 1. (Let μ_n denote the group of nth roots of 1, so $\mu_n \subseteq k$.) If $a \in k^*$, the Kummer theory allows us to manufacture a character χ_a of the Galois group of the extension $k(a^{1/n})/k$. Let us recall this construction. The Galois group $G(k(a^{1/n})/k)$ is cyclic, and if $\alpha = a^{1/n}$ and σ belongs to this Galois group, then $\sigma(\alpha)/\alpha$ is an nth root of 1. So the element $\frac{\sigma(\alpha)}{\alpha} \in \mu_n$, and if ζ is a *chosen* primitive n^{th} root of 1 (in k), then ζ gives an isomorphism of μ_n with Z/nZ. Using this isomorphism, we transfer $\frac{\sigma(\alpha)}{\alpha}$ to Z/nZ, and then we set

$$\chi_a(\sigma) = \frac{1}{n}\frac{\sigma(\alpha)}{\alpha} \in \frac{1}{n}\mathbf{Z}/\mathbf{Z} \subseteq \mathbf{Q}/\mathbf{Z} .$$

The function χ_a is the required character. This association yields a pairing of k^*/k^{*n} and the group G_k/G_k^n into $\mathbf{Q}/\mathbf{Z}$ (the Kummer pairing) which is an exact duality of finite groups (k is local!).

Now the natural choice of notation yields a symbol $<a, b> = <\chi_a, b> = b \cup \delta\chi_a \in Br(k)$ —but this depends upon the choice of ζ, and even applying inv_k does not change this. It is a marvelous fact that if we go back *via* ζ to the nth roots of 1 after applying inv_k, then the dependence on ζ disappears. This is

PROPOSITION 59. *Let* k *be a local field and let* $a, b \in k^*$. *Define the (multiplicative) Hilbert Symbol as follows: Let* ζ *be a chosen primitive* n^{th} *root of* 1, *and set*

$$(a, b) = \zeta^{n \cdot inv_k <a, b>} = \zeta^{\{a, b\}} ,$$

where $<a, b>$ *is defined as above via Kummer Theory. Then* (a,b) *is an* n^{th} *root of unity which is independent of the choice of* ζ, *it depends only upon* a *and* b. *In fact,* (a, b) *can be computed by the formula*

$$(a, b) = \frac{\theta_k(b)(\alpha)}{\alpha} , \quad \textit{where } \alpha^n = a .$$

Proof: Everything will follow from the formula

$$(a, b) = \frac{\theta_k(b)(\alpha)}{\alpha} .$$

But
$$(a, b) = \zeta^{n\, inv_k <a, b>} = \zeta^{n\, inv_k(b \cup \delta\chi_a)}$$
$$= \zeta^{n\, \chi_a(\theta_k(b))}$$
$$= \frac{\theta_k(b)(\alpha)}{\alpha}, \text{ by definition of } \chi_a. \qquad \text{Q.E.D.}$$

PROPOSITION 60. *The Hilbert Symbol* (a, b) *has the following properties:*

(1) $(a_1 a_2, b) = (a_1, b)(a_2, b)$

(2) $(a, b_1 b_2) = (a, b_1)(a, b_2)$

(3) $(a, b) = 1 \Longleftrightarrow b \in \mathfrak{N}_{L/k} L^*$, *where* $L = k(\alpha)$, *and* $\alpha^n = a$.

(4) $(a, -a) = (a, 1-a) = 1$

(5) $(a, b) = (b, a)^{-1}$

(6) *If* $(a,b) = 1$ *for all* $b \in k^*$, *then* $a \in k^{*n}$.

Proof: The first two properties are trivial consequences of the bilinearity of cup products. If b is a norm, then $(b, L/k) = 1$, and by Proposition 59, we deduce that $(a, b) = 1$. If $(a, b) = 1$, then $\theta_k(b)$ fixes α, and α generates the field extension $L = k(\alpha)$. Hence, $(b, L/k) = 1$; and the reciprocity law shows that b is a norm from L. Let us show that (4) implies (5). We have

$$\begin{aligned} 1 &= (ab, -ab) = (a, -ab)(b, -ab) \\ &= (a, -a)(a, b)(b, a)(b, -b) \\ &= (a, b)(b, a), \text{ which is (5).} \end{aligned}$$

To prove (4), factor the polynomial $X^n - a$ in $L = k(\alpha)$. We obtain

$$X^n - a = \prod_{j=0}^{n-1} (X - \zeta^j \alpha) .$$

Now if $\alpha^m \in k$ for some $m \leqq n$, then $m|n$ (by Kummer Theory) and L is cyclic of degree m over k. In this case, the $k-$ conjugates of $X - \zeta^j \alpha$ are exactly the terms $X - \zeta^\ell \alpha$, where $\ell \equiv j \bmod (n/m)$; hence,

$$X^n - a = \prod_{\ell=0}^{d-1} \mathfrak{N}_{L/k}(X - \zeta^\ell \alpha) = \mathfrak{N}_{L/k}\Big(\prod_{\ell=0}^{d-1} X - \zeta^\ell \alpha\Big) .$$

When $X = 0$ or 1 we obtain property (4) by using property (3).

Lastly, if $(a, b) = 1$ for all $b \in k^*$, then $\theta_k(b)\alpha = \alpha$ for all $b \in k^*$.

But the image of k^* is dense in G_k, so we see that G_k fixes α. Thus, $\alpha \in k^*$ and $a = \alpha^n \in k^{*n}$. Q.E.D.

As a corollary, we shall deduce Proposition 56 for primes distinct from ch(k).

COROLLARY. *If an element* $b \in k^*$ *is a norm from every cyclic extension of degree* p $(\neq \mathrm{ch}(k))$, *then* $b \in k^{*p}$.

Proof: For each $a \in k^*$, property (3) yields $(a, b) = 1$. Property (5) implies $(b, a) = 1$ for each $a \in k^*$, and property (6) completes the proof. Q.E.D.

The above corollary settles the existence theorem for local fields of characteristic zero, i.e., the p-adic fields. We must now treat the formal power series fields of characteristic $p > 0$. The methods are analogous—we use Artin-Schreier Theory to replace Kummer Theory, and since we do not introduce Witt vectors, we must explicitly calculate the local symbol we obtain.

So now assume $k = k_0((t))$—the field of formal power series in one variable over the finite field k_0. Let $p = \mathrm{ch}(k) = \mathrm{ch}(k_0)$, then the pth roots of 1 lie in k. If $a \in k^*$, the Artin-Schreier theory allows us to manufacture a character χ_a of the Galois group G_k (really of the Galois group of the cyclic extension of degree p given by $k(\wp^{-1}(a))$, where $\wp(\alpha) = \alpha^p - \alpha$ and $\wp^{-1}(a)$ means any element α with $\wp(\alpha) = a$). Let us recall this construction. If $\alpha = \wp^{-1}(a)$ and σ belongs to the Galois group of $k(\alpha)$ over k, then $\sigma\alpha - \alpha$ is in the prime field, i.e., lies in $\mathbf{Z}/p\mathbf{Z}$. We set

$$\chi_a(\sigma) = \frac{1}{p}(\sigma(\alpha) - \alpha) \in \frac{1}{p}\mathbf{Z}/\mathbf{Z} \subseteq \mathbf{Q}/\mathbf{Z},$$

and the function χ_a is the required character. This association yields a pairing of $k^+/\wp(k^+)$ and the group G_k/G_k^p into $\mathbf{Q}/\mathbf{Z}$ (the Artin-Schreier pairing) which is an exact duality of topological groups.

Now the natural choice of notation yields a symbol

$$\{a, b> = <\chi_a, b> = b \cup \delta\chi_a \in \mathrm{Br}(k)$$

and if we apply inv_k, we get an element of $\mathbf{Q}/\mathbf{Z}$. To interpret this element properly in terms of a, b and k, we need it to have values in $\mathbf{Z}/p\mathbf{Z}$ – the prime field. With this in mind we have

PROPOSITION 61. *Let* k *be a local field of characteristic* $p > 0$, *and let* $a \in k$ *and* $b \in k^*$. *Define the (semi-additive) Hilbert Symbol (or, Artin- Schreier symbol) as follows:*

$$[a, b) = p \cdot \mathrm{inv}_k \{a, b> ,$$

where $\{a, b>$ *is defined as above via Artin-Schreier Theory. Then* $[a, b)$ *can be computed by the formula*

$$[a, b) = \theta_k(b)(\alpha) - \alpha, \text{ where } \wp(\alpha) = a.$$

Proof: By definition,

$$[a, b) = p \cdot \mathrm{inv}_k(b \cup \delta\chi_a) = p\chi_a(\theta_k(b))$$
$$= p \cdot \frac{1}{p}(\theta_k(b)\alpha - \alpha) = \theta_k(b)(\alpha) - \alpha . \qquad \text{Q.E.D.}$$

Observe that $[a, b) \in \mathbf{Z}/p\mathbf{Z}$ for every choice of a, b. If $a \in k$, then a dt is a "meromorphic" differential form, and we may define its residue as the coefficient of the term dt/t. This is independent of the choice of the uniformizing parameter t (see [SA] for the explicit proof), and so we get an element res (a dt) $\in k_0$.

PROPOSITION 62. *The Artin-Schreier Symbol* $[a, b)$ *has the following properties:*

(1) $[a_1 + a_2, b) = [a_1, b) + [a_2, b)$

(2) $[a, b_1 b_2) = [a, b_1) + [a, b_2)$

(3) $[a, b) = 0 \iff b \in \mathfrak{N}_{L/k} L^*$, *where* $L = k(\alpha)$, *and* $\wp(\alpha) = a$

(4) $[a, a) = 0$, *for all* $a \in k^*$

(5) *If* $[a, b) = 0$, *for all* $b \in k^*$, *then* $a \in \wp(k)$.

Proof: The first two properties are trivial consequences of the bilinearity of cup products. If b is a norm, then $(b, L/k) = 1$, and by Proposition 61, we deduce that $[a, b) = 0$. Conversely, if $[a, b) = 0$, then $(b, L/k)$ fixes α; hence it fixes L. Therefore $(b, L/k) = 1$, and b is a norm from L. To prove (4), observe that if $a \in \wp(k)$, say $a = \wp(\alpha)$ with $\alpha \in k$, then Proposition 61 shows that $[a, a) = 0$. So suppose that $a \notin \wp(k)$, then $L = k(\alpha)$ with $\wp(\alpha) = a$ is a cyclic extension of degree p, and the conjugates of α are $(\alpha - j)$ where $j = 0, 1, \ldots, p-1$. Thus,

$$\mathfrak{N}_{L/k}(\alpha) = \prod_{j=0}^{p-1} (\alpha - j) = \wp(\alpha) = a \ ,$$

and property (3) applies. If $[a, b) = 0$ for all $b \in k^*$, then $\theta_k(b)\alpha = \alpha$ for all b, and (as before) the $\theta_k(b)$ are dense in G_k. Thus $\alpha \in k$, and $a \in \wp(k)$. Q.E.D.

PROPOSITION 63 (H. L. Schmid[Sd]). *If* $k = k_0((t))$ *and* $a \in k$, *and* $b \in k^*$, *then*

$$[a, b) = \mathrm{tr}_{k_0/\mathbf{Z}/p\mathbf{Z}}(\mathrm{res}(a \frac{db}{b})) \ .$$

Proof: We first show that if $c = \mathrm{res}(a \frac{db}{b})$, then $[a, b) = [c, t)$. Now any $b \in k^*$ has the form $b = b_0 t^r$, where b_0 is a unit of k^*; so using the bilinearity of our pairing, we see that it will suffice to prove the result when $\mathrm{ord}_k(b) = 1$. But then b is a uniformizing parameter, and the residue is independent of the choice of such, so we may assume $b = t$ at the outset. Write a as a power series in t

$$a = \sum_n a_n t^n = \sum_{n<0} a_n t^n + a_0 + \sum_{n>0} a_n t^n \ .$$

Again by the additivity of both functions of a, $[a, t)$ and $[\mathrm{res}(a\frac{dt}{t}), t)$, we see that we may treat each summand separately. The middle summand yields the tautology $[a_0, t) = [a_0, t)$. The summand $\Sigma_{n>0}\, a_n t^n$ is always in $\wp(k)$. In fact, if

$$a = \sum_{n>0} a_n t^n \ , \quad \text{let} \quad \alpha = \sum_{j=0}^{\infty} a^{p^j}$$

then $\wp(\alpha) = \alpha^p - \alpha = -a$. Proposition 61 shows that $[a, t)$ is trivial, and checking shows that $c = \operatorname{res}(a\frac{dt}{t}) = 0$.

There remains the most interesting case, $a = \Sigma_{n<0}\, a_n t^n$. In this case, our "functions" a being "meromorphic," the sum is finite. It follows from the bilinearity that we may even assume $a = a_n t^n$ for some fixed $n < 0$. Now $\operatorname{res}(a\frac{dt}{t}) = 0$, so we must show that $[a, t) = 0$. Property (2) of Proposition 62 shows that $[a_{-n}t^{-n}, t^{-n}) = -n[a_{-n}t^{-n}, t) = -n[a, t)$. But

$$[a_{-n}t^{-n}, a_{-n}t^{-n}) = [a_{-n}t^{-n}, a_{-n}) + [a_{-n}t^{-n}, t^{-n}) \ ;$$

hence by (4) of Proposition 62,

$$0 = [a_{-n}t^{-n}, a_{-n}) - n[a, t) \ .$$

However, k_0 being finite (hence perfect), $a_{-n} = x^p$, so $[a, a_{-n}) = [a, x^p) = p[a, x) = 0$; and we deduce that $-n[a, t) = 0$. If $n \not\equiv 0 \pmod p$, this proves that $[a, t) = 0$.

Suppose then that $n = mp$. As before, write $a_{-n} = x^p$, then $a = (xt^{-m})^p = \wp(xt^{-m}) + xt^{-m}$. We obtain

$$[a, t) = [\wp(xt^{-m}), t) + [xt^{-m}, t) = [xt^{-m}, t) \ .$$

If $m \not\equiv 0 \pmod p$ we are done; if $m \equiv 0 \pmod p$, we repeat. The obvious induction yields $[a, t) = 0$, as required.

It is now simple to prove Schmid's formula. Our argument shows that $[a, b) = [\operatorname{res}(a\frac{db}{b}), t)$; therefore, we need only prove that if $c \in k_0$ we have

$$[c, t) = \operatorname{tr}_{k_0/\mathbf{Z}/p\mathbf{Z}}(c) \ .$$

Look at the extension of k_0 given by $k_1 = k_0(\gamma)$ where $\wp(\gamma) = c$. Set $L = k_1((t))$, then L is an unramified extension of k, and the determination of the reciprocity law in unramified extensions (Proposition 51) yields

$(t, L/k) = F$ —the Frobenius. We now apply Proposition 62, and we get

$$\begin{aligned}[c, t) &= F(\gamma)-\gamma = \gamma^q - \gamma \qquad (q = |k_0|)\\ &= (\gamma^q - \gamma^{q/p}) + (\gamma^{q/p} - \gamma^{q/p^2}) + \cdots + (\gamma^p - \gamma)\\ &= c^{q/p} + c^{q/p^2} + \cdots + c\\ &= \mathrm{tr}_{k_0/\mathbf{Z}/p\mathbf{Z}}(c) \qquad \text{Q.E.D.}\end{aligned}$$

COROLLARY. *If* $k = k_0((t))$, *with* k_0 *a finite field of characteristic* $p > 0$, *then an element* $b \in k^*$ *which is a norm from every cyclic extension of* k *of degree* p *is a* p^{th} *power.*

Proof: Property (3) of Proposition 62 shows that $[a, b) = 0$ for all $a \in k$. Were b *not* a pth power, the differential db/b would not vanish. Given $c \in k_0$, we would then be able to find elements $a \in k$, such that $\mathrm{res}(a\frac{db}{b}) = c$. By Schmid's formula, $\mathrm{tr}(c) = [a, b) = 0$, and this contradicts the fact that the trace: $k_0^+ \to \mathbf{Z}/p\mathbf{Z}$ is surjective. Q.E.D.

Remarks. 1) The corollary above finally completes the proof of the existence theorem of §4.

2) The reader will find explicit calculations of the Hilbert Symbol in [Hl, Ha, He]; especially noteworthy is the application to the quadratic reciprocity law—historically the first known instance of Artin's Reciprocity Law.

3) We have taken some care to point out the analogies between (a, b) and $[a, b)$. These stem from the fact that both are cup-products of suitable 1-dimensional cohomology groups into $\mathbf{Q}/\mathbf{Z}$. Now (a, b) gives a duality of k^*/k^{*n} with itself by Kummer Theory and the reciprocity law, while $[a, b)$ yields a duality of $k^+/\wp(k^+)$ with k^*/k^{*p} (the former discrete, the latter compact) by Artin-Schreier Theory and the reciprocity law. There are other dualities of the same nature, the last basic one being that between k^+/k^{+p} and itself $(\mathrm{ch}(k) = p > 0)$ *via* the formula

$$[x, y] = \mathrm{tr}_{k_0/\mathbf{Z}/p\mathbf{Z}}(\mathrm{res}(x\,dy)) .$$

This is also a cup-product of suitable one-dimensional cohomology groups, and the whole subject of these local symbols takes on a coherence it did not formerly possess when viewed in this cohomological light. We shall treat such matters in the next chapter.

CHAPTER VI

DUALITY

We connect profinite group theory, arithmetic (as represented by local class field theory), and algebraic geometry. The results will center about the duality of certain cohomology groups, and this duality will reflect the local class field theory. Algebraic geometry will enter when we show how the cohomology of profinite groups (at least in its arithmetic applications) can be subsumed in a more general cohomology procedure currently used in algebraic geometry.

§1. Dualizing Modules and Poincaré Groups

The following result, due to Grothendieck [GF], is fundamental for what we intend to do.

PROPOSITION 64 (Grothendieck). *Let* $\mathfrak{C}$ *be a noetherian, abelian category, and let* T *be a contravariant functor from* $\mathfrak{C}$ *to* Ab *(the category of abelian groups). Then a necessary and sufficient condition that* T *be ind-representable (that is, representable by an object* I *of* dir lim $\mathfrak{C}$*) is that* T *be left-exact.*

Proof: Necessity is clear, so assume T is left-exact. We say a pair $\langle A, x\rangle$ consisting of an object A of $\mathfrak{C}$ and an element $x \in T(A)$ is *minimal* iff for each proper quotient, B, of A (i.e., $A \to B \to 0$ is exact, but $A \neq B$) the element x does NOT belong to T(B). (Observe that as T is left-exact, the sequence $0 \to T(B) \to T(A)$ is exact.) We partially order the minimal pairs as follows: $\langle A, x\rangle \succeq \langle A', x'\rangle$ if and only if there is a morphism $A' \to A$, say ϕ, such that $T(\phi)$ carries x to x′. If such a ϕ

exists, it must be unique. For if two such morphisms ϕ, ψ were to exist, we could form the cokernel of the pair of morphisms $A' \rightrightarrows A$, say C. As T is left-exact, we would get an exact sequence

$$0 \to T(C) \to T(A) \underset{T(\psi)}{\overset{T(\phi)}{\rightrightarrows}} T(A') ,$$

and this would show that $x \in T(C)$, a contradiction.

The ordering $\geqq$ has the Moore-Smith property because $<A \amalg A', <x, x'>>$ is a common dominant for $<A, x>$, $<A', x'>$. Set $I =$ dir lim A, the limit being taken over the minimal pairs $<A, x>$ with the ordering $\geqq$. Define T(I) by $T(I) = \text{proj lim } T(A)$, then there is a canonical element $\xi \in T(I)$, namely the element obtained from all the $x \in T(A)$.

Our element ξ defines a homomorphism of functors

$$\theta : \text{Hom}_{\mathfrak{C}}(-, I) \to T$$

via the rule:

$$\theta(u) = T(u)(\xi) \in T(A), \text{ for } u \in \text{Hom}_{\mathfrak{C}}(A, I) .$$

Were θ not injective, we could pick a maximal counterexample, say A_0 ($\mathfrak{C}$ is noetherian!). Thus, there would exist $u \neq 0$, $u \in \text{Hom}_{\mathfrak{C}}(A_0, I)$ with $\theta(u) = 0$. If $B = A_0 \amalg A$ where A is an object of $\mathfrak{C}$ part of a minimal pair, and if $v \in \text{Hom}_{\mathfrak{C}}(A, I)$ is the canonical morphism, then as A_0 is a proper subobject of B, the mapping $\theta : \text{Hom}_{\mathfrak{C}}(B, I) \to T(B)$ is injective. However, $\theta(u, v) = \theta(0, v) = <0, x>$ where x is the canonical object of T(A) – a contradiction. In a similar way one proves that θ is surjective as well, and the proof is complete. Q.E.D.

To apply this Proposition to the cohomology of profinite groups, let G be a profinite group, and let $\mathfrak{C}$ be the category of finite G-modules. If H is an abelian group, we let H^d be its Pontrjagin dual. (Thus $H^d =$ Hom (H, R/Z) (= Hom (H, Q/Z) if H is a torsion group).) We give H^d the compact-open topology, then one knows that H^d is compact when H is discrete, finite when H is finite, etc.

PROPOSITION 65. *Let* G *be a profinite group with* $\mathrm{cd}\,G \leqq n$. *Assume* $H^n(G, A)$ *is finite for all finite* G-*modules* A. *Then the functor* $A \rightsquigarrow H^n(G, A)^d$ *is representable; in fact, the representing object is a torsion* G-*module*.

Proof: If $T(A) = H^n(G, A)^d$ then as $\mathrm{cd}\,G \leqq n$, T is left-exact. The category $\mathfrak{C}$ is noetherian, so Grothendieck's result applies. We get an object I of dir lim $\mathfrak{C}$ which represents T. But, dir lim $\mathfrak{C}$ is the category of torsion G-modules; so we are done. Q.E.D.

The G-module I is called the *dualizing module* (in dimension n) for G and it has the usual uniqueness properties for objects representing functors. Observe that T(I) is the compact group which is the dual of the discrete cohomology group $H^n(G, I)$. The canonical element $\xi \in T(I)$ is thereby a *homomorphism* $\xi \colon H^n(G, I) \to \mathbf{Q}/\mathbf{Z}$.

Now Proposition 65 says that $\mathrm{Hom}_G(A, I)$ is dual to $H^n(G, A)$, for any finite G-module A. If we let $A^{(d)} = \mathrm{Hom}_{\mathbf{Z}}(A, I)$ and refer to $A^{(d)}$ as the "I-dual" of A, then it is easy to make $A^{(d)}$ into a G-module in such a way that $H^0(G, A^{(d)}) = \mathrm{Hom}_G(A, I)$. We simply define (σf) by the formula:

$$(\sigma f)(a) = \sigma(f(\sigma^{-1} a)) \quad .$$

When this is done we see that Proposition 65 yields a duality between $H^0(G, A^{(d)})$ and $H^n(G, A)$. If we pass to the direct limit of such A, then the $A^{(d)}$ are replaced by a projective limit, and when this is done $H^0(G, A^{(d)})$ is a compact group in its natural topology. It is the Pontrjagin dual of the discrete torsion group $H^n(G, A)$. We have therefore proved the following Corollary to Proposition 65.

COROLLARY. *If* A *is a torsion* G-*module and if* $A^{(d)}$ *is its* I-*dual, then the compact group* $H^0(G, A^{(d)})$ *and the discrete group* $H^n(G, A)$ *are Pontrjagin duals, one of the other.*

Let us observe one more thing about dualizing modules for a profinite group G. *If* H *is an open subgroup of* G *and if* I *is the dualizing module* *for* G *then it is the dualizing module for* H. To prove this, we have to show that $H^n(H,A)$ is dual to $H^0(H, A^{(d)})$ for all finite H-modules A. By Shapiro's lemma, $H^n(H, A) = H^n(G, \pi_* A)$, and as I is the dualizing module for G, we obtain the duality

$$H^n(H, A) \text{ dual to } H^0(G, (\pi_* A)^{(d)}) \ .$$

Now the canonical injection $0 \to A \to \pi_* A$ yields a functorial mapping $\mathrm{Hom}_G(\pi_* A, I) \to \mathrm{Hom}_H(A, I)$ which one checks is an isomorphism. Therefore, $H^n(H, A)$ is dual to $H^0(H, A^{(d)})$, as required.

Obviously, if we are only interested in p-primary components of the cohomology or in p-primary modules, we need only assume $\mathrm{cd}_p G \leqq n$. In this vein, Proposition 65 yields a wonderful criterion that $\mathrm{s\,cd}_p G = \mathrm{cd}_p G + 1$.

PROPOSITION 66 (Serre). *Let* G *be a profinite group of* p*-cohomological dimension* n *and assume the other hypotheses of Proposition 65 are valid. Let* I *be the dualizing module for* G, *then* $\mathrm{s\,cd}_p G = n + 1$ *if and only if* G *has an open subgroup* U *such that* I^U *has a subgroup isomorphic to the* p*-primary part of* $\mathbf{Q}/\mathbf{Z}$.

Proof: $(\mathbf{Q}/\mathbf{Z}\,; p)$ is contained in I^U if and only if $\mathrm{Hom}_U((\mathbf{Q}/\mathbf{Z}; p), I) \neq (0)$. Since I is dualizing, this means exactly that $H^n(U, \mathbf{Q}/\mathbf{Z}\,; p)$ shall not vanish. But $H^n(U, \mathbf{Q}/\mathbf{Z}) \xrightarrow{\sim} H^{n+1}(U, \mathbf{Z})$; hence, $(\mathbf{Q}/\mathbf{Z})_p \subseteq I^U \Longleftrightarrow H^{n+1}(U, \mathbf{Z}; p) \neq (0)$. And now, Proposition 18 completes the proof. Q.E.D.

We turn now to those profinite p-groups which exhibit the phenomenon of "Poincaré duality" familiar from topology. We shall say that a profinite p-group is *an* n*-dimensional Poincaré group* if G possesses the following three properties:

(a) $H^r(G, \mathbf{Z}/p\mathbf{Z})$ *is a finite group for all* r,

(b) $\dim_{\mathbf{Z}/p\mathbf{Z}} H^n(G, \mathbf{Z}/p\mathbf{Z}) = 1$,

(c) *The pairing* $\mathbf{Z}/p\mathbf{Z} \times \mathbf{Z}/p\mathbf{Z} \to \mathbf{Z}/p\mathbf{Z}$ *induces a perfect duality of vector spaces via the cup-product*

$$H^r(G, \mathbf{Z}/p\mathbf{Z}) \times H^{n-r}(G, \mathbf{Z}/p\mathbf{Z}) \to H^n(G, \mathbf{Z}/p\mathbf{Z}), \quad r \geqq 0.$$

Of course, we assume $n \geqq 1$. According to (c), $H^m(G, Z/pZ)$ for $m > r$ must vanish, so we deduce that *an* n*-dimensional Poincaré Group has* $cd_p = n$. When $n = 1$, a 1-dimensional Poincaré Group must be a free profinite p-group of rank 1 by our results on generators and relations in profinite p-groups. When n is arbitrary the existence of plenty of n-dimensional Poincaré groups is given by the following result of M. Lazard (unpublished, unfortunately): *If* G *is an* n*-dimensional Lie Group over the field of* p*-adic numbers, then all sufficiently small open subgroups of* G *are* n*-dimensional Poincaré Groups.*

Since every finite p-primary G-module has a composition series whose factors are simple, finite p-primary G-modules, and since the only finite, simple, p-primary G-module for a profinite p-group is Z/pZ (Proposition 17), we deduce that $H^r(G, A)$ is a finite group for each finite G-module. A. We have shown $cd\,G = n$ when G is an n-dimensional Poincaré group, so we may apply Proposition 65 and obtain a dualizing module I for G. Now we will see that the canonical pairing $A \times A^{(d)} \to I$ induces (*via* cup-product) the *duality pairings* $H^r(G, A) \times H^{n-r}(G, A^{(d)}) \to Q/Z$ for each finite module A and each $r \in Z$. This is the exact analog of classical Poincaré Duality. For this result we will need:

PROPOSITION 67. *Every open subgroup of an* n*-dimensional Poincaré group is an* n*-dimensional Poincaré group. If* U *is such an open subgroup of* G*, then the mapping*

$$\mathrm{tr}:\ H^n(U, Z/pZ) \to H^n(G, Z/pZ)$$

is an isomorphism. If A *is a finite* G*-module, then* proj lim $H^r(U, A)$ *vanishes for* $r \neq n$, *where the projective limit is taken over all open subgroups of* G *with respect to the transfer mappings. For* $r = n$, *the functor* $A \rightsquigarrow \operatorname{proj\,lim}_U H^n(U, A)$ *is exact.*

Proof: Let U be a given open subgroup of G, and form $A = \pi_{*_{U \to G}}(Z/pZ)$. This is a G-module which is killed by p, so it is a vector space over Z/pZ. Its dual, A^*, is paired with it into Z/pZ.

This induces cup-product pairings

$$H^r(G, A) \times H^{n-r}(G, A^*) \to H^n(G, Z/pZ)$$

We claim these pairings are perfect dualities of finite dimensional spaces. When A is Z/pZ (i.e., when U is taken to be G itself), this is part (c) of the definition of Poincaré group. We shall therefore use induction on dim A. If A is given, we write

$$0 \to A' \to A \to A'' = Z/pZ \to 0$$

for a subspace A' (such exist as the only simple G-module of p-power order is Z/pZ). The dual sequence

$$0 \to A''^* \to A^* \to A'^* \to 0$$

and cohomology yield a diagram

$$\begin{array}{ccccccccc}
\cdots \longrightarrow & H^{r+1}(A')^d & \longrightarrow & H^r(A'')^d & \longrightarrow & H^r(A)^d & \longrightarrow & H^r(A')^d & \longrightarrow \cdots \\
& \downarrow \theta' & & \downarrow \theta'' & & \downarrow \theta & & \downarrow \theta' & \\
\cdots \longrightarrow & H^{s-1}(A'^*) & \longrightarrow & H^s(A''^*) & \longrightarrow & H^s(A^*) & \longrightarrow & H^s(A'^*) & \longrightarrow \cdots
\end{array}$$

where the vertical arrows are induced by our cup-product, and where $H^j(A)$ means $H^j(G, A)$, etc. (Of course, $r + s = n$.) The above diagram commutes or anti-commutes in each square, and the induction assumption shows that the maps θ', θ'' are isomorphisms. The five-lemma now yields our claim.

Now Shapiro's Lemma applied to the case $A = \pi_{*_{U \to G}}(Z/pZ)$, shows that the cup-product pairing

$$H^r(U, Z/pZ) \times H^{n-r}(U, Z/pZ) \to H^n(U, Z/pZ)$$

is a perfect duality. For $r = n$, we deduce that $H^n(U, Z/pZ)$ is of dimension 1 over Z/pZ.

To show that tr is an isomorphism, it suffices to show that it is surjective. However, there is an inclusion $\mathrm{Hom}_G(A, I) \hookrightarrow \mathrm{Hom}_U(A, I)$ for every A; so by duality there is a *surjection* $H^n(U, A) \to H^n(G, A)$. The latter map is exactly the transfer (as one easily sees), and the result follows from taking $A = Z/pZ$.

By the Mittag-Leffler criterion ([GD]), the functor proj lim is exact on the category of profinite (hence on finite) G-modules. It follows that the functors $A \rightsquigarrow \operatorname{proj\,lim}_U H^r(U, A)$ form a δ-functor. By arguing with a composition series, if we wish to prove these functors trivial, it suffices to do so for $A = Z/pZ$. But in this case, the duality just proved shows that what we must prove is $\operatorname{dir\,lim}_U H^{n-r}(U, Z/pZ) = (0)$, for $r \neq n$. This is true (and trivial) for all profinite groups G and all modules A, let alone Z/pZ.

Since $\operatorname{proj\,lim} H^r(U, A) = (0)$ for $r < n$, the functor $\operatorname{proj\,lim}_U H^n(U, A)$ is left-exact. However, $cd_p G \leqq n$ because G is Poincaré; so $\operatorname{proj\,lim}_U H^n(U, A)$ is right-exact as well. Q.E.D.

THEOREM 40. *Let* G *be an* n-*dimensional Poincaré Group with dualizing module* I. *Then*

(a) I *is isomorphic to* $(Q/Z; p)$ *as an abelian group,*

(b) *The canonical homomorphism* $\xi: H^n(G, I) \to Q/Z$ *is a monomorphism whose image is* $(Q/Z; p)$,

(c) *For every finite,* p-*primary,* G-*module* A, *the cup-product pairings*

$$H^r(G, A) \times H^{n-r}(G, A^{(d)}) \to H^n(G, I) \hookrightarrow Q/Z$$

are perfect dualities of finite abelian groups for every $r \in Z$.

Proof: For the moment assume conclusions (a) and (b); we shall now show how (c) follows from these assumed conclusions. The proof of (c) is by induction on the length of a composition series for the given module A. We observe that $A \rightsquigarrow A^{(d)}$ is an exact functor because I is divisible by (a), and we further observe that for simple A (i.e., $A = Z/pZ$) $A^{(d)}$ is naturally isomorphic to the vector space dual of A, A^*. By the definition of Poincaré groups, conclusion (c) is true in the special case: A is simple. For arbitrary A, we find a maximal submodule A', so that $A'' = A/A'$ is simple, and we get two exact sequences

$$0 \to A' \longrightarrow A \longrightarrow A'' \longrightarrow 0$$
$$0 \to A''^{(d)} \longrightarrow A^{(d)} \longrightarrow A'^{(d)} \longrightarrow 0$$

Our pairings induce the commutative (or anti-commutative) diagram of the corresponding cohomology sequences

$$\begin{array}{ccccccccc}
\cdots \to & H^r(G, A')^d & \longrightarrow & H^{r-1}(G, A'')^d & \longrightarrow & H^{r-1}(G, A)^d & \longrightarrow & H^{r-1}(G, A')^d & \longrightarrow \cdots \\
& \downarrow \theta'_{r,s} & & \downarrow \theta''_{r-1,s+1} & & \downarrow \theta_{r-1,s+1} & & \downarrow & \\
\cdots \to & H^s(G, A'^{(d)}) & \longrightarrow & H^{s+1}(G, A''^{(d)}) & \longrightarrow & H^{s+1}(G, A^{(d)}) & \longrightarrow & H^{s+1}(G, A'^{(d)}) & \longrightarrow \cdots
\end{array}$$

By induction hypothesis the maps $\theta'_{r,s}, \theta''_{u,v}$ for $r+s = u+v = n$ are all isomorphisms; so by the five lemma, $\theta_{r,s}$ is an isomorphism for all r, s with $r+s = n$.

We must therefore prove assertions (a), (b). Now if A is annihilated by p, then A is a finite dimensional vector space over Z/pZ, and precisely the same argument as in Proposition 67 yields:

The cup-product pairings

$$H^r(G, A) \times H^{n-r}(G, A^*) \longrightarrow H^n(G, Z/pZ)$$

are exact dualities of finite dimensional vector spaces for every $r \in Z$ *and every* G*-module* A *annihilated by* p. (We do not need (a) or (b) for this proof.)

If we apply this statement to the case $r = n$, we see that $H^n(G, A)$ is the vector space dual of $H^0(G, A^*)$, or what is the same, $H^n(G, A)^d$ is functorially isomorphic to $H^0(G, A^*) = \mathrm{Hom}_G(A, Z/pZ)$. However, by the definition of dualizing modules, $H^n(G, A)^d$ is also functorially isomorphic to $\mathrm{Hom}_G(A, I) = \mathrm{Hom}_G(A, I_p)$. Therefore both Z/pZ and I_p represent the functor $A \leadsto H^n(G, A)^d$ on the category G-modules annihilated by p. Thus $I_p = Z/pZ$. Since I is a p-primary torsion group, *it follows that* I *is either* Z/p^sZ *or* $(Q/Z; p)$ *as abelian group.*

We can now apply Proposition 67. According to this proposition, proj lim $H^n(U, A)$ is an exact functor of A. However, because I is the dualizing module for each such U, we deduce that

$$H^n(U, A) \text{ is dual to } \mathrm{Hom}_U(A, I) \ .$$

Thus, the functor $\operatorname{dir\,lim}_U \mathrm{Hom}_U(A, I) = \mathrm{Hom}\,(A, I)$ is an exact functor of A. This proves that I is injective as abelian group; hence $I = (\mathbf{Q}/\mathbf{Z}\,;\, p)$, which proves (a).

Since (a) holds, the $\mathbf{Z}$-endomorphisms of I are exactly the elements of the ring of p-adic integers, $\mathbf{Z}_p$. But the action of G commutes with the action of these endomorphisms in view of the simple nature in which $\mathbf{Z}_p$ acts on I. Hence, $\mathrm{Hom}_G(I, I) = \mathbf{Z}_p$, also. However, I is the dualizing module; so $\mathrm{Hom}_G(I, I)$ is dual to $H^n(G, I)$. Thus, there is a canonical isomorphism of $H^n(G, I)$ with the dual of $\mathbf{Z}_p$, that is, with $(\mathbf{Q}/\mathbf{Z}\,;\, p)$. But the element ξ corresponds to the element $\mathrm{id} \in \mathrm{Hom}_G(I, I) = T(I)$; hence ξ is the above canonical isomorphism, and (b) is proved. Q.E.D.

Observe that (a) implies $(A^{(d)})^{(d)} \xrightarrow{\sim} A$ as a G-module, so our duality is as close as possible to perfection. Also, as $\mathrm{End}\,(I) = \mathbf{Z}_p$, we see that $\mathrm{Aut}\,(I)$ = group of units, U_k, of the p-adic field $k = \mathbf{Q}_p$. Since G acts on I by automorphisms, we find that there is a canonical homomorphism (called χ by Serre)

$$\chi : G \to U_k \ .$$

As G is a p-group, the image of χ lies in U_1 (see Ch. V, §1). We can reformulate Serre's criterion on strict cohomological dimension in terms of χ: *In order that* $\mathrm{scd}\, G = n + 1$, *it is necessary and sufficient that* $\mathrm{Im}\,\chi$ *be finite*. To prove this observe that χ being continuous (because G acts *continuously on* I), a necessary and sufficient condition that $\mathrm{Im}\,\chi$ be finite is that χ vanish on some open subgroup, say U, of G. Thus U acts trivially on I, so $I^U = I = (\mathbf{Q}/\mathbf{Z}\,;\, p)$ —and Serre's criterion applies.

The reader will find many other results in Serre's notes [SG] – especially important is the letter of Tate reproduced in these notes. Here, one finds a strengthened form of Proposition 64: *Let* C(R) *be the category of* R-*modules for a topological ring* R *whose topology is given by letting the two-sided ideals form a fundamental system of neighborhoods of zero, and let* $C_0(R)$ *denote the subcategory of* C(R) *consisting of those modules* M *which are the unions of their submodules* $M_{\mathfrak{A}} = \mathrm{Hom}_R(R/\mathfrak{A}, M)$. *If* T *is an additive contravariant functor from* $C_0(R)$ *to* Ab, *transforming* dir lim *into* proj lim, *then* T *is representable if and only if* T *is left-exact.* One also finds strengthened duality theorems, and as a consequence of these theorems one may deduce: *If a profinite* p*-group* G *has finite cohomological dimension, then* G *is a Poincaré group if and only if at least one of its open subgroups is a Poincaré group.* If we put this statement together with Lazard's result, we get: *Every analytic pro-*p*-group of finite cohomological dimension is a Poincaré group.*

One can also find an approach to the duality theorems very much in keeping with present ideas on the subject in the appendix to Serre's notes. This appendix, due to Verdier, introduces dualizing complexes and other machinery now used in algebraic geometry in questions of duality.

§2. Grothendieck Topologies and Cohomology

We will explain a very general method for constructing cohomology groups currently used in algebraic geometry. Galois cohomology is a very special case of this theory. The idea is to construct a general sheaf theory and take cohomology groups with coefficients in a sheaf. For sheaf theory one does not need the whole theory of topological spaces, merely the notion of an open covering—it is the latter notion which we abstract. The basic idea and theory is entirely due to Grothendieck, in fact, the germ of this idea communicated to Tate lead him to develop the profinite cohomology.

Start with a category $\mathfrak{C}$ —whose objects should be thought of as the "open sets" in our general topology. Given an object T of $\mathfrak{C}$, we consider a family of morphisms $\phi_\alpha : U_\alpha \to T$ in $\mathfrak{C}$ (denoted $\{U_\alpha \to T\}$), and we isolate a collection of such families for various T. Denote this collection by Cov($\mathfrak{T}$), and let $\mathfrak{T}$ denote the *pair* consisting of the category $\mathfrak{C}$ and its collection Cov($\mathfrak{T}$). The data described in $\mathfrak{T}$ will be called a *Grothendieck*

Topology (with covering families the elements of Cov $\mathfrak{T}$) if and only if the following three axioms are satisfied

GT 1) If $U \xrightarrow[\phi]{} V$ is an isomorphism then $\{U \to V\} \in \operatorname{Cov} \mathfrak{T}$.

GT 2) If $\{T_\alpha \longrightarrow T\} \in \operatorname{Cov}\mathfrak{T}$ and if $\{U_{\beta\alpha} \to T_\alpha\}_\beta \in \operatorname{Cov}\mathfrak{T}$ for all α, then $\{U_{\beta\alpha} \to T\}_{\beta\alpha}$ belongs to Cov $\mathfrak{T}$. (That is: "An open covering of an open covering is an open covering.")

GT 3) If $\{T_\alpha \to T\} \in \operatorname{Cov} T$ and V is an arbitrary object of $\mathfrak{C}$ *over* T (i.e., $\exists \phi: V \to T$), then the fibred products $T_\alpha \times_T V$ exist for all α, and $\{T_\alpha \times_T V \to V\} \in \operatorname{Cov} T$.

Remarks. In this theory fibred products play the role that intersection does in ordinary topology. For example, the simplest case is that arising from ordinary topology as follows: If X is a topological space, let $\mathfrak{C}$ be the category whose objects are the open sets of X and whose morphisms are given by

$$\operatorname{Hom}_{\mathfrak{C}}(U, V) = \begin{Bmatrix} \emptyset, & \text{if } U \nsubseteq V \\ \text{the inclusion map, if } U \subseteq V \end{Bmatrix}.$$

Define Cov $\mathfrak{T}$ by: $\{U_\alpha \to U\} \in \operatorname{Cov}\mathfrak{T}$ iff $\bigcup_\alpha U_\alpha = U$.

The data just described yield a Grothendieck topology (the reader should check that $T_\alpha \times_T V = T_\alpha \cap V$, for example), let us call it the *classical Grothendieck topology.*

If $<\mathfrak{C}, \operatorname{Cov}\mathfrak{T}>$ is a Grothendieck topology, then a contravariant functor from $\mathfrak{C}$ to Ab is called a *presheaf of abelian groups on* $\mathfrak{T}$. The presheaves form an abelian category $\mathfrak{P}_\mathfrak{T}$, in the obvious way. The category $\mathfrak{P}_\mathfrak{T}$ has generators and satisfies AB5 [GT], so it possesses sufficiently many injectives.

More precisely, a sequence $F' \to F \to F''$ is exact in $\mathfrak{P}_\mathfrak{T}$ iff for all $T \in \operatorname{Ob}\mathfrak{C}$, $F'(T) \to F(T) \to F''(T)$ is exact in the usual sense. If $\mathfrak{C}'$ is another category and if $f: \mathfrak{C} \to \mathfrak{C}'$ is a functor then for every presheaf F on $\mathfrak{C}'$, the functor $F \circ f = f^{\mathfrak{P}}(F)$ is a presheaf on $\mathfrak{C}$. So

$$f^{\mathfrak{P}}(F)(T) = F(f(T)).$$

By Kan's process [K], we can construct a functor $f_{\mathfrak{P}}$ which is left adjoint to $f^{\mathfrak{P}}$; the functor $f_{\mathfrak{P}}$ is right exact while the functor $f^{\mathfrak{P}}$ is exact. Let us recall the idea of the proof. Given $T' \in \mathrm{Ob}\,\mathfrak{C}'$, let $I^f_{T'}$ be the category whose objects are pairs $<U, \phi>$ such that $U \in \mathrm{Ob}\,\mathfrak{C}$ and ϕ maps $T' \to f(U)$ in $\mathfrak{C}'$, and whose morphisms are those $v : U \to \tilde{U}$ so that the diagram

$$\begin{array}{ccc} & T' & \\ {\scriptstyle\phi}\swarrow & & \searrow{\scriptstyle\tilde{\phi}} \\ f(U) & \xrightarrow[f(v)]{} & f(\tilde{U}) \end{array}$$

commutes. We set $(f_{\mathfrak{P}} F)(T') = \varinjlim_{I^f_{T'}} F(U)$. Then one checks that $f_{\mathfrak{P}} F$ is a presheaf, that $f_{\mathfrak{P}}$ is a functor, and that

$$(*) \qquad \mathrm{Hom}_{\mathfrak{C}}(F, f^{\mathfrak{P}} G) \xrightarrow{\sim} \mathrm{Hom}_{\mathfrak{C}'}(f_{\mathfrak{P}} F, G) \ .$$

We can use $f_{\mathfrak{P}}$ to describe the generators of $\mathfrak{P}_{\mathcal{T}}$. If $X \in \mathrm{Ob}\,\mathfrak{C}$, let $\{X\}$ be the category whose objects consist just of X, and whose morphisms are just 1_X. $\mathfrak{P}_{\{X\}} = \mathrm{Ab}$ as is clear, and there is the inclusion functor $i : \{X\} \to \mathfrak{C}$. Thus, for all $F \in \mathfrak{P}_{\{X\}}$ we have $i_{\mathfrak{P}} F \in \mathfrak{P}_{\mathcal{T}}$; moreover, for all $G \in P_{\mathcal{T}}$, one evidently finds that $i^{\mathfrak{P}} G = G(X)$. Now it is easy to see that $i_{\mathfrak{P}}$ is exact; hence $i^{\mathfrak{P}}$ carries injectives to injectives (use equation (*)). Moreover, if we set $\mathcal{Z}_X = i_{\mathfrak{P}} \mathbf{Z}$, then by (*) again,

$$\mathrm{Hom}_{\mathcal{T}}(\mathcal{Z}_X, G) = \mathrm{Hom}_{\{X\}}(\mathbf{Z}, G(X)) = G(X) \ ;$$

so $\mathcal{Z}_X$ represents the functor $G \longrightarrow G(X)$ from $\mathfrak{P}_{\mathcal{T}}$ to Ab. The family $\{\mathcal{Z}_X \mid X \in \mathrm{Ob}\,\mathfrak{C}\}$ is a family of generators for the category $\mathfrak{P}_{\mathcal{T}}$.

Sheaves can now be introduced as special types of presheaves. If F is a presheaf on $\mathcal{T}$, we say that F *is a sheaf on* $\mathcal{T}$ *iff the sequence*

$$(S) \qquad F(T) \xrightarrow{\theta} \prod_{\alpha} F(T_\alpha) \underset{\theta_2}{\overset{\theta_1}{\rightrightarrows}} \prod_{\alpha,\beta} F(T_\alpha \times_T T_\beta)$$

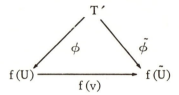

is exact for every covering $\{T_\alpha \to T\}$ *in* $\mathcal{T}$. Thus, we require that θ map $F(T)$ *bijectively* onto the set of all $\xi = (\xi_\alpha) \in \Pi_\alpha F(T_\alpha)$ such that $\theta_1(\xi) = \theta_2(\xi)$. The sheaves form a category by decreeing that it shall be a *full* subcategory of $\mathfrak{P}_{\mathcal{T}}$. Denote the category of sheaves on $\mathcal{T}$ by $\mathfrak{S}_{\mathcal{T}}$. We have the inclusion functor

$$i : \mathfrak{S}_{\mathcal{T}} \hookrightarrow \mathfrak{P}_{\mathcal{T}}$$

which regards each sheaf just as a presheaf. It is a basic result of Grothendieck (see [AG] for the proof) that i possesses an adjoint, $\# : \mathfrak{P}_{\mathcal{T}} \to \mathfrak{S}_{\mathcal{T}}$. If $F \in \mathfrak{P}_{\mathcal{T}}$, let $\#(F)$ be denoted $F^{\#}$, it is called the *sheaf associated to* F. We have the adjointness property

$$(**) \qquad \mathrm{Hom}_{\mathfrak{P}_{\mathcal{T}}}(F, iG) \cong \mathrm{Hom}_{\mathfrak{S}_{\mathcal{T}}}(F^{\#}, G) \quad .$$

From these facts one finds that $\mathfrak{S}_{\mathcal{T}}$ is an abelian category with generators (the $\mathcal{Z}_X^{\#} = \mathbf{Z}_X$ are generators) and AB5, it has products, i is left exact, and # is exact. Hence, $\mathfrak{S}_{\mathcal{T}}$ possesses enough injectives, and the functor $\Gamma_X : \mathfrak{S}_{\mathcal{T}} \to \mathrm{Ab}$ *via* $\Gamma_X(F) = F(X)$ is left exact. In fact, $\Gamma_X(F) = \mathrm{Hom}_{\mathfrak{S}_{\mathcal{T}}}(\mathbf{Z}_X, F)$.

We can now introduce cohomology. First the cohomology for presheaves—the Čech cohomology. Choose some $V \in \mathcal{O}b\, \mathfrak{C}$, and some covering $\{V_\alpha \to V\}$. We may then form $V_\alpha \times_V V_\beta$, $V_\alpha \times_V V_\beta \times_V V_\gamma, \ldots, V_{\alpha_1} \times_V \cdots {}_V\times V_{\alpha_r}, \ldots$. These give rise to families of maps

$$V \leftarrow \{V_\alpha\}_\alpha \leftleftarrows \{V_\alpha \times_V V_\beta\}_{\alpha\beta} \;\substack{\longleftarrow\\ \longleftarrow\\ \longleftarrow}\; \{V_\alpha \times_V V_\beta \times_V V_\gamma\}_{\alpha\beta\gamma} \;\substack{\longleftarrow\\ \longleftarrow\\ \longleftarrow\\ \longleftarrow}\; \cdots \; .$$

For a presheaf F on $\mathcal{T}$, we obtain a diagram of maps and groups

$$F(V) \longrightarrow \Pi_\alpha F(V_\alpha) \rightrightarrows \Pi_{\alpha\beta} F(V_\alpha \times_V V_\beta) \;\substack{\longrightarrow\\ \longrightarrow\\ \longrightarrow}\; \Pi_{\alpha\beta\gamma} F(V_\alpha \times_V V_\beta \times_V V_\gamma)$$

$$\cdots \;\substack{\longrightarrow\\ \vdots\\ \longrightarrow}\; \Pi_{\alpha_0, \ldots, \alpha_r} F(V_{\alpha_0} \times_V \cdots \times_V V_{\alpha_r}) \;\substack{\longrightarrow\\ \vdots\\ \longrightarrow}\; \cdots \; .$$

Let $C^r(\{V_\alpha \to V\}, F)$ be the group $\Pi_{\alpha_0,\ldots,\alpha_r} F(V_{\alpha_0} \times_V \cdots \times_V V_{\alpha_r})$, and let $d^r: C^r(\{V_\alpha \to V\}, F) \to C^{r+1}(\{V_\alpha \to V\}, F)$ be the alternating sum of the $r+2$ maps from C^r to C^{r+1} as above. (This sum is taken in the group C^{r+1}.) We obtain a sequence

$$F(V) \to C^0(\{V_\alpha \to V\}, F) \xrightarrow{d^0} C^1(\{V_\alpha \to V\}, F) \xrightarrow{d^1} \cdots$$

$$\cdots \xrightarrow{d^{r-1}} C^r(\{V_\alpha \to V\}, F) \xrightarrow{d^r} C^{r+1}(\{V_\alpha \to V\}, F) \to \cdots$$

which is easily seen to be a *complex*. We shall call this complex the *Čech cochain complex of the covering* $\{V_\alpha \to V\}$ *with coefficients in* F. Its cohomology $H^r(\{V_\alpha \to V\}, F)$ will be called the *Čech cohomology of the covering* $\{V_\alpha \to V\}$ *with coefficients in* F. (Here, $H^0(\{V_\alpha \to V\}, F) = \ker d^0$.) Observe that

$$H^0(\{V_\alpha \to V\}, F) = F(V), \text{ if } F \in \mathfrak{S}_{\mathcal{T}} .$$

It is important to know that the functors $H^q(\{V_\alpha \to V\}, F)$ are effacable for all $q > 0$. For then it will follow that $\{H^q(\{V_\alpha \to V\}, -)\}_q$ is a universal δ-functor from the category $\mathfrak{P}_{\mathcal{T}}$ to Ab; hence $H^q(\{V_\alpha \to V\}, -)$ is the qth derived functor of $H^0(\{V_\alpha \to V\}, -)$.

PROPOSITION 68. *The functors* $H^q(\{V_\alpha \to V\}, -)$ *are effacable for every* $q > 0$.

Proof: (Grothendieck) It suffices to show that $H^q(\{V_\alpha \to V\}, F) = (0)$ for an injective presheaf F, and all $q > 0$. This is the same as showing that the complex

$$\Pi_\alpha F(V_\alpha) \to \Pi_{\alpha,\beta} F(V_\alpha \times_V V_\beta) \to \Pi_{\alpha\beta\gamma} F(V_\alpha \times_V V_\beta \times_V V_\gamma) \to \cdots$$

is an exact sequence. Now we know that $\mathrm{Hom}(\mathcal{Z}_{V_\alpha}, F) = F(V_\alpha)$, etc.; hence, we need to show that

$$\Pi_\alpha \mathrm{Hom}(\mathcal{Z}_{V_\alpha}, F) \to \Pi_{\alpha\beta} \mathrm{Hom}(\mathcal{Z}_{V_\alpha \times_V V_\beta}, F) \to \cdots$$

is exact. But the latter sequence is exactly

$$\operatorname{Hom}(\coprod_{\alpha} \mathcal{Z}_{V_\alpha}, F) \to \operatorname{Hom}(\coprod_{\alpha\beta} \mathcal{Z}_{V_\alpha \times_V V_\beta}, F) \to \operatorname{Hom}(\coprod_{\alpha\beta\gamma} \mathcal{Z}_{V_\alpha \times V_\beta \times V_\gamma}, F) \to \cdots$$

and, as F is injective, its exactness is equivalent with the exactness of

$$\coprod_{\alpha} \mathcal{Z}_{V_\alpha} \longleftarrow \coprod_{\alpha\beta} \mathcal{Z}_{V_\alpha \times V_\beta} \longleftarrow \coprod_{\alpha\beta\gamma} \mathcal{Z}_{V_\alpha \times V_\beta \times V_\gamma} \longleftarrow \cdots .$$

By definition this means that we need to check the exactness of

$$\coprod_{\alpha} \mathcal{Z}_{V_\alpha}(X) \longleftarrow \coprod_{\alpha\beta} \mathcal{Z}_{V_\alpha \times V_\beta}(X) \longleftarrow \coprod_{\alpha\beta\gamma} \mathcal{Z}_{V_\alpha \times V_\beta \times V_\gamma}(X) \longleftarrow \cdots$$

for every object X of $\mathfrak{C}$.

From the definition of $\mathcal{Z}_V$ and the explicit construction of $i_{\mathfrak{P}}Z$ as a limit, one can check that $\mathcal{Z}_V(X) = \coprod_{\operatorname{Hom}(X,V)} Z$. This implies that the last sequence is induced by the following diagram of the indexing sets:

$$\coprod_{\alpha} \operatorname{Hom}(X, V_\alpha) \leftleftarrows \coprod_{\alpha\beta} \operatorname{Hom}(X, V_\alpha \times V_\beta) \overset{\longleftarrow}{\underset{\longleftarrow}{\longleftarrow}} \coprod_{\alpha\beta\gamma} \operatorname{Hom}(X, V_\alpha \times V_\beta \times V_\gamma) .$$

Certain maps are common to all terms in this diagram, namely those in Hom (X, V); hence "factoring this set out" we get the diagram of indexing sets

$$\coprod_{s \in \operatorname{Hom}(X,V)} \left(\coprod_{\alpha} \operatorname{Hom}_s(X, V_\alpha) \leftleftarrows \coprod_{\alpha\beta} \operatorname{Hom}_s(X, V_\alpha \times V_\beta) \overset{\longleftarrow}{\underset{\longleftarrow}{\longleftarrow}} \cdots \right) .$$

Here, $\operatorname{Hom}_s(X, V_\alpha)$ means those maps $X \to V_\alpha$ which render the diagram

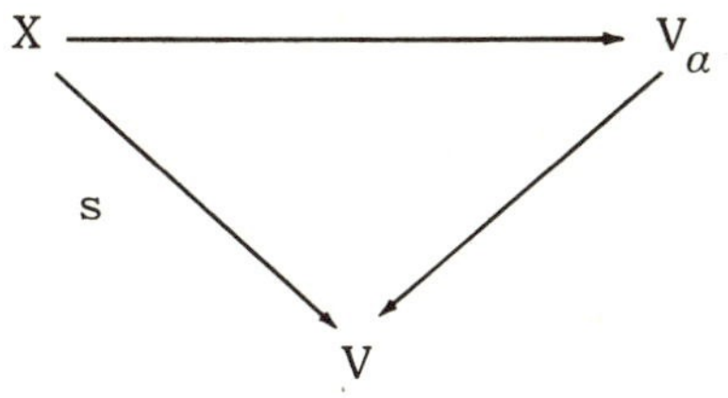

commutative. On the other hand, if $M_s = \operatorname{Hom}_s(X, V_\alpha)$, then clearly

$$M_s \times M_s = \mathrm{Hom}_s(X, V_\alpha \times V_\beta), \quad \text{etc.};$$

so, we get the diagram of indexing sets

$$\coprod_{s \in \mathrm{Hom}(X,V)} \left(M_s \leftleftarrows M_s \times M_s \;\substack{\longleftarrow\\\longleftarrow\\\longleftarrow}\; M_s \times M_s \times M_s \;\substack{\longleftarrow\\\longleftarrow\\\longleftarrow\\\longleftarrow}\; \cdots \right)$$

If we put this information together, we see that our complex (whose exactness we need to prove) becomes a sum of complexes, each of the form

$$\coprod_M Z \leftleftarrows \coprod_{M \times M} Z \;\substack{\longleftarrow\\\longleftarrow\\\longleftarrow}\; \coprod_{M \times M \times M} Z \;\substack{\longleftarrow\\\longleftarrow\\\longleftarrow\\\longleftarrow}\; \cdots .$$

But such a complex possesses a contracting homotopy [CE], because if $n_{i_0, \ldots, i_m}$ is an integer in the $< i_0, \ldots, i_m >$ component of $\coprod_{M^{m+1}} Z$, and if ξ is a fixed element of M, then

$$n_{i_0, \ldots, i_m} \longrightarrow n_{\xi, i_0, \ldots, i_m}$$

is the required contracting homotopy. Thus our complexes are acyclic, as required. Q.E.D.

Suppose now that $\{V_\alpha \to V\}$ and $\{W_\beta \to V\}$ are two coverings of V. Let I be the index set for the covering $\{V_\alpha \to V\}$, let J be that for $\{W_\beta \to V\}$. Let us say that $\{V_\alpha \to V\}$ is *bigger than or equal to* $\{W_\beta \to V\}$ if there is a map $I \to J$ (say h), and there is a morphism $V_\alpha \to W_{h(\alpha)}$ over V for all α. (In the classical case this just means that $\{V_\alpha \to V\}$ *refines* $\{W_\beta \to V\}$.) We write $\{V_\alpha \to V\} \geqq \{W_\beta \to V\}$ in this case.

It is obvious that if $\{V_\alpha \to V\} \geqq \{W_\beta \to V\}$, then there is an induced map of cohomology groups

$$H^q(\{W_\beta \to V\}, F) \longrightarrow H^q(\{V_\alpha \to V\}, F) \;;$$

and the usual proof (see [Go]) shows that the latter cohomology map is independent of the choice of map from the covering $\{V_\alpha \to V\}$ to the covering $\{W_\beta \to V\}$. Hence, the coverings of V form a partially ordered set under $\geqq$, and the groups $H^q(\{V_\alpha \to V\}, F)$ form a directed system on this indexing set. We define *the Čech cohomology groups of* V *with coefficients in* F, denoted $\check{H}^q(V, F)$, by the formula

$$\operatorname*{dir\,lim}_{\text{(coverings of } V)} H^q(\{V_\alpha \to V\}, F) = \check{H}^q(V, F), \qquad q \geqq 0.$$

Clearly, Proposition 68 shows that $\{\check{H}^q(V, -)\}_q$ is a universal δ-functor from $\mathfrak{P}_{\mathcal{T}}$ to Ab; so the groups $\check{H}^q(V, F)$ are derived functors of $\check{H}^0(V, F)$.

The cohomology for sheaves is on the one hand easier to define and on the other more difficult to compute than that for presheaves. If V is an object of $\mathfrak{C}$, then the functor

$$F \rightsquigarrow \Gamma_V(F) = F(V)$$

is a left exact functor from $\mathfrak{S}_{\mathcal{T}}$ to Ab. Its right derived functors $R^q\Gamma_V(F)$ exist (since $\mathfrak{S}_{\mathcal{T}}$ possesses enough injectives), and these are the *cohomology groups of* V *with coefficients in* F. So, we obtain

$$H^q(V, F) = R^q\Gamma_V(F); \quad H^0(V, F) = F(V) .$$

Observe that if F is a sheaf, $\check{H}^0(V, F) = F(V)$.

The functor $i: \mathfrak{S}_{\mathcal{T}} \to \mathfrak{P}_{\mathcal{T}}$ is left-exact, as we know; hence, it possesses right derived functors $R^q i$. The standard notation for these right derived functors is $\mathcal{H}^q(\ )$; so $R^q i(F) = \mathcal{H}^q(F)$. For each *sheaf* F, we get a *presheaf* $\mathcal{H}^q(F)$. By definition, $\mathcal{H}^0(F) = F$, and I claim that

$$\mathcal{H}^q(F)(U) = H^q(U, F).$$

To see this, observe that $U \rightsquigarrow H^q(U, F)$ is a presheaf, it agrees with $\mathcal{H}^q(F)(U)$ when $q = 0$. Therefore, both $\{\mathcal{H}^q(\ )\}$, and $\{U \rightsquigarrow H^q(U, -)\}$ are universal δ-functors from $\mathfrak{S}_{\mathcal{T}}$ to $\mathfrak{P}_{\mathcal{T}}$, and the claim is proved.

The principal use of the functors $\mathcal{H}^q(F)$ is to obtain a spectral sequence relating Čech cohomology and sheaf cohomology. If $\{V_\alpha \to V\} \in \operatorname{Cov} \mathcal{T}$, and if F is a sheaf, then $F(V) = H^0(\{V_\alpha \to V\}, F)$. But, $H^0(\{V_\alpha \to V\}, F)$ is the composition of the functor $i: \mathfrak{S}_{\mathcal{T}} \to \mathfrak{P}_{\mathcal{T}}$ with the functor

$H^0(\{V_\alpha \to V\}, -)$. Therefore,

$$\Gamma_V = H^0(\{V_\alpha \to V\}, -) \circ i ,$$

and since i takes injectives to injectives, there is a spectral sequence of composed functors

$$R^p H^0(\{V_\alpha \to V\}, -) \circ R^q i \Longrightarrow R^* \Gamma_V .$$

Plugging in the variable F, we obtain *the spectral sequence*

$$\text{(SS1)} \qquad H^p(\{V_\alpha \to V\}, \mathcal{H}^q(F)) \Longrightarrow H^*(V, F) .$$

Passing (SS1) to the limit over the coverings of V, we obtain the *spectral sequence of Čech cohomology*

$$\text{(SS2)} \qquad \check{H}^p(V, \mathcal{H}^q(F)) \Longrightarrow H^*(V, F) .$$

The sequence of terms of low degree for (SS2) is

$$0 \to \check{H}^1(V, F) \to H^1(V, F) \to \check{H}^0(V, \mathcal{H}^1(F)) \to \check{H}^2(V, F) \to H^2(V, F).$$

In this low degree exact sequence, the term $\check{H}^0(V, \mathcal{H}^1(F))$ vanishes. This is a consequence of

PROPOSITION 69. *For each* $q > 0$, *the groups*

$$\check{H}^0(V, \mathcal{H}^q(F))$$

vanish.

Proof: $\mathcal{H}^q(F)$ is a presheaf for each q, and one sees that if $\mathcal{H}^q(F)^\#$ denotes the associated sheaf, then

$$\check{H}^0(V, \mathcal{H}^q(F)) \subseteq \mathcal{H}^q(F)^\#(V) .$$

(This is a consequence of the construction of #, see [AG].) Therefore, it will suffice to prove $\mathcal{H}^q(F)^\#(V) = (0)$ for each V and each $q > 0$. But $\# \circ i = 1_{\mathcal{S}}$, so the spectral sequence for composed functors yields

$$R^p \# \circ R^q i \Longrightarrow R^* 1_{\mathcal{S}} .$$

Both $1_{\mathcal{S}}$ and # are exact, so their higher derived functors vanish; and it follows immediately that $\mathcal{H}^q(F)^{\#} = (0)$ for $q > 0$. Q.E.D.

COROLLARY. *For each sheaf* F *and each object* V *of* $\mathfrak{C}$, *we have*

$$\check{H}^1(V,F) \xrightarrow{\sim} H^1(V,F)$$
$$\check{H}^2(V,F) \subseteq H^2(V,F)\ .$$

The preceeding theory has been condensed, abstract, without illustrative examples, and probably mildly incomprehensible. In seeking to rectify this, we shall now show how profinite group cohomology fits into this new setting. Galois cohomology and its generalizations will be treated in the next section.

Let G be a profinite group, let $\mathfrak{C}$ be the category of finite G-sets (with *continuous* G-action), and let Cov $\mathcal{T}$ be the collection of finite families of maps $\{V_\alpha \to V\}$ such that $\bigcup_\alpha \mathrm{Im}(V_\alpha) = V$ (i.e., the family $\{V_\alpha \to V\}$ is surjective).

We obtain a Grothendieck topology, $\mathcal{T}$, which we can call *the* G-*topology*, and denote $\mathcal{T}_G$.

THEOREM 41. *Let* G *be a profinite group and let* $\mathcal{T}_G$ *be the* G-*topology described above. If* $\mathcal{S}_G$ *denotes the category of sheaves on* $\mathcal{T}_G$, *then* $\mathcal{S}_G$ *is equivalent to the category of* G-*modules (with continuous action). This equivalence is given by*

$$F \rightsquigarrow \varinjlim_{U \text{ open in } G} F(G/U) = A_F$$

where G/U *is the left-coset space of the open subgroup* U, *and the limit is taken over all open subgroups of* G. *For the* G-*module*, A_F, *we have*

$$A_F^U = F(G/U)\ ;$$

so if $\{1\}$ *denotes the one-point set (with trivial* G-*action), then the groups* $H^r(\{1\}, F)$ *are precisely the cohomology groups* $H^r(G, A_F)$ *for the module*

A_F. *In addition, spectral sequence* (SS2) *degenerates, so that cohomology may be computed in all dimensions by the Čech method—which is the method of cochains, cocycles, etc. introduced at the beginning. Spectral sequence* (SS1) *becomes the Hochschild-Serre sequence in the special case of the covering* $\{G/N \to \{1\}\}$, *where* $N \lhd G$.

Proof: The category, $\mathfrak{C}$, of finite G-sets possesses cokernel pairs and quotients and is noetherian. If one examines the proof of Proposition 64, one sees that this is all we need to obtain its conclusion. A sheaf is just a left exact, contravariant functor from $\mathfrak{C}$ to Ab; hence by Proposition 64, every sheaf is ind – representable. The representing object A_F is therefore a direct limit of finite G-sets; so A_F is a G-set. As F has values in Ab, A_F is a group object in the category of G-sets; that is, A_F is a G-module. Since A_F is the direct limit of *finite* G-sets (each of which has *continuous* G-action), we see that G acts continuously on A_F. Conversely, every continuous G-module M gives a sheaf on $\mathfrak{C}$ by the formula $F_M(S) = \mathrm{Hom}_G(S, M)$; this proves that $\mathcal{S}_G$ is equivalent to the category of G-modules.

To get the formula stated in the theorem, we compute dir lim F (G/U).

$$\operatorname*{dir\,lim}_U F(G/U) = \operatorname*{dir\,lim}_U \mathrm{Hom}_G(G/U, A_F) = \mathrm{Hom}_G(\operatorname*{proj\,lim}_U G/U, A_F)$$

$$= \mathrm{Hom}_G(G, A_F) = A_F \ .$$

Since $F(G/U) = \mathrm{Hom}_G(G/U, A_F)$, an element of $F(G/U)$ is precisely an element of A_F which is fixed by U. Therefore, $F(\{1\}) = A_F^G = H^0(G, A_F)$; so the universal δ-functors

$$\begin{array}{ccc} F & \rightsquigarrow & \{H^r(\{1\}, F)\}_r \\ \| & & \\ A_F & \rightsquigarrow & \{H^r(G, A_F)\}_r \end{array}$$

agree when $r = 0$.

We must now show that spectral sequence (SS2) degnerates. Given $p \geqq 0$, and $q > 0$, we need to prove $\check{H}^p(\{1\}, \mathcal{H}^q(F)) = (0)$. We have done the case $p = 0$ in general, therefore we assume $p > 0$. Now the coverings $\{G/U \to \{1\}\}$ are final among all coverings of $\{1\}$ in the category $\mathfrak{C}$; hence

$$\check{H}^p(\{1\}, \mathcal{H}^q(F)) = \operatorname*{dir\,lim}_U H^p(\{G/U \to \{1\}\}, \mathcal{H}^q(F)) \ .$$

From this equation, we see that if the assertion concerning (SS1) is true, then

$$\begin{aligned} H^p(\{1\}, \mathcal{H}^q(F)) &= \operatorname*{dir\,lim}_U H^p(G/U, H^q(U, A_F)) \\ &= H^p(\operatorname{proj\,lim} G/U, \operatorname{dir\,lim} H^q(U, A_F)) \ . \end{aligned}$$

But, $\operatorname*{dir\,lim}_U H^q(U, A_F)$ (under restriction, clearly) is always zero (for $q > 0$), so the assertion for (SS2) is now proved modulo that for (SS1). We now must compute $H^p(\{G/U \to \{1\}\}, \mathcal{H}^q(F))$. By definition, it is the defect from exactness in the diagram

$$H^q_{\mathcal{J}}((G/U)^p, F) \ \substack{\longrightarrow \\ \vdots \\ \longrightarrow} \ H^q_{\mathcal{J}}((G/U)^{p+1}, F) \ \substack{\longrightarrow \\ \vdots \\ \longrightarrow} \ H^q_{\mathcal{J}}((G/U)^{p+2}, F) \ ,$$

where the subscript $\mathcal{J}$ is meant as a reminder that the above groups are Grothendieck cohomology groups. The G-set $(G/U)^p$ splits into $|(G/U)|^{p-1}$ disjoint copies of G/U (as G-set); hence we get the diagram

$$\prod_{|G/U|^{p-1}} H^q_{\mathcal{J}}(G/U, F) \ \substack{\longrightarrow \\ \vdots \\ \longrightarrow} \ \prod_{|G/U|^{p}} H^q_{\mathcal{J}}(G/U, F) \ \substack{\longrightarrow \\ \vdots \\ \longrightarrow} \ \prod_{|G/U|^{p+1}} H^q_{\mathcal{J}}(G/U, F) \ .$$

But $H^q_{\mathcal{J}}(G/U, F)$ is the qth derived functor of $F \rightsquigarrow F(G/U)$; that is, it is the qth derived functor of $A_F \rightsquigarrow A_F^U$. Hence, we obtain the diagram

$$\prod_{|G/U|^{p-1}} H^q(U, A_F) \ \substack{\longrightarrow \\ \vdots \\ \longrightarrow} \ \prod_{|G/U|^{p}} H^q(U, A_F) \ \substack{\longrightarrow \\ \vdots \\ \longrightarrow} \ \prod_{|G/U|^{p+1}} H^q(U, A_F) \ ,$$

or put more conventionally, the diagram

$$C^{p-1}(G/U, H^q(U, A_F)) \to C^p(G/U, H^q(U, A_F)) \to C^{p+1}(G/U, H^q(U, A_F)) \ .$$

This immediately shows that $H^p(\{G/U \to \{1\}\}, \mathcal{H}^q(F))$ is the group $H^p(G/U, H^q(U, A_F))$, as required. Q.E.D.

§3. Galois, Étale, and Flat Cohomology

We wish to use the theory of §2 directly without the intermediary of Theorem 41; in fact there will be important cases where profinite groups are inadequate but the general method still works.

Let X be a scheme, we shall define two topologies "on X". The first topology is the *étale topology on* X, its category $\mathfrak{C}_e$ is the category of all schemes over X of finite presentation (even locally so) which are *étale* over X (i.e., flat and unramified, see [GD_2]). Covering families are families $\{U_\alpha \to U\}$ whose image in U is all of U, i.e., the scheme $\coprod_\alpha U_\alpha$ is *faithfully flat* over U, see [GD_2]. Denote the *étale* topology over X by $\mathcal{T}_e(X)$ or $\mathcal{T}_e$ if reference to X is clear. The second topology is the *flat topology* on X.(*) Its category $\mathfrak{C}_f$ is the category of all schemes over X (locally) of finite presentation, quasi-finite, and flat over X. (See [GD_2] for definitions of these terms.) The coverings are just as in the *étale* case (only this time we must assume the flatness of all maps $U_\alpha \to U$, it is not automatic). Denote the flat topology over X by $\mathcal{T}_f(X)$ or $\mathcal{T}_f$ if reference to X is clear.

The corresponding categories of sheaves are denoted $\mathcal{S}_e$ (or $\mathcal{S}_e(X)$), $\mathcal{S}_f$ (or $\mathcal{S}_f(X)$) respectively. Cohomology for the *étale* topology consists in the derived functors of $F \rightsquigarrow F(X)$ for $F \in \mathcal{S}_e$; similarly for flat cohomology. The notations $H_e^r(X, F)$, $H_f^r(X, F)$ separate the two possible definitions of cohomology.

In all of what follows, X will always be an affine scheme (even the spec of a field!) so we will use the notation $H^r(A, F)$ as a substitute for $H^r(X, F)$ when $X = \operatorname{Spec} A$. If $\{U_\alpha \to X\}$ is a covering of the affine scheme $X = \operatorname{Spec} A$, then the characterization of affine schemes shows that our covering factors through the covering $\{V_\alpha \to X\}$ where $V_\alpha = \operatorname{Spec}(\Gamma(U_\alpha, \mathcal{O}_{U_\alpha}))$. For this reason, when dealing with coverings of affine schemes, we shall always treat the case in which the covering schemes are themselves affine.

(*) We adopt a slightly simplified flat topology here, quasi-finiteness is not always assumed in the general case.

Since affine schemes are (quasi) compact, any large covering family $\{U_\alpha \to X = \operatorname{Spec} A\}$ is dominated by a finite subfamily; therefore we may and do assume that all covering families are finite.

Let us then take some affine $Y = \operatorname{Spec} A$, and a finite affine covering $\{Y_\alpha \to Y\}_\alpha$. By setting $Y' = \coprod_\alpha Y_\alpha$, we obtain a covering with one morphism, and Y' is affine, say $Y' = \operatorname{Spec} B$. The Čech cohomology for the covering $\{Y' \to Y\}$ is then entirely dependent upon the algebra "extension" B/A — we shall call it the *Čech cohomology in the layer* B/A. If we translate to affine rings, we see that presheaves are covariant functors on $\mathfrak{C}^0$ (where $\mathfrak{C}^0$ is the category of $\Gamma(X, \mathcal{O}_X) = A(X)$-algebras), and

$$\underbrace{Y' \times \cdots \times Y'}_{\text{r-times}} = \operatorname{Spec}(\otimes_A^r B).$$

So by definition, the Čech cohomology in the layer B/A (denoted $H^r(B/A, F)$) is the cohomology of the complex

$$F(B) \to F(B \otimes_A B) \to F(\otimes_A^3 B) \to F(\otimes_A^4 B) \to \cdots .$$

To say that the covering $\{Y'' \to Y\}$ is greater than or equal to the covering $\{Y' \to Y\}$ means merely that the diagram of A-algebras

$$\begin{array}{ccc} A(Y') = B & \dashrightarrow & C = A(Y'') \\ & \nwarrow \quad \nearrow & \\ & A = A(Y) & \end{array}$$

may be completed to a commutative diagram by the addition of an arrow $B \to C$ as shown. If we are given a direct mapping family of such A-algebras, say $\{A_\alpha\}$, then we get a system of "finer and finer" coverings $\{A_\alpha \to A\}$, and we *define* $H^r(\operatorname{dir\,lim} A_\alpha/A, F)$ by the formula

$$H^r(\operatorname{dir\,lim} A_\alpha/A, F) = \operatorname*{dir\,lim}_\alpha H^r(A_\alpha/A, F) .$$

If the category of A-algebras possesses a final object (in the sense of coverings!), say $\bar{A}$, then clearly

$$H^r(\bar{A}/A, F) = \operatorname*{dir\,lim}_{\text{all coverings}} H^r(A_\alpha/A, F) = \check{H}^r(A, F) .$$

Should F commute with direct limits, then $H^r(\bar{A}/A, F)$ is merely the cohomology of the complex

$$F(\bar{A}) \to F(\bar{A} \otimes_A \bar{A}) \to \cdots \to F(\otimes_A^n \bar{A}) \to \cdots .$$

For example, this will hold if F is a sheaf. In particular, if F is representable, then the representing object will be a commutative group scheme and the above will hold. If G is this group scheme

$$(\text{so that } F(C) = \text{Hom}(\text{Spec } C, G))$$

we write $H^r(B/A, G)$, $\check{H}^r(A, G)$, $H^r(\bar{A}/A, G)$ instead of $H^r(B/A, F)$, etc.

One important case in which the A-algebras possess a final object (for coverings) is the case in which $A = k$ is a field. The final object is then the algebraic closure of k in the flat case and the separable closure in the *étale* case. (An *étale* extension of k is a product of *separable* field extensions, while quasi-finite and flat allow products of Artin local rings with residue fields finite over k.) Thus for the case of fields k, we deduce the important formulas

$$(*)\qquad \begin{aligned} \check{H}^r_e(k, F) &= H^r(k_s/k, F) = \operatorname*{dir\,lim}_{K/k,\ \text{sep.}} H^r(K/k, F) \\ \check{H}^r_f(k, F) &= H^r(\bar{k}/k, F) = \operatorname*{dir\,lim}_{K/k,\ \text{alg}} H^r(K/k, F); \end{aligned}$$

and we recall that these groups may be computed as the cohomology of

$$\begin{aligned} &F(k_s) \to F(k_s \otimes_k k_s) \to \cdots \\ &F(\bar{k}) \to F(\bar{k} \otimes_k \bar{k}) \to \cdots \end{aligned}$$

whenever F is a sheaf (or more generally commutes with dir lim).

The notation used in formulas (*) above suggests earlier notation for Galois cohomology. This is not accidental, for we have

PROPOSITION 70. *Let* F *be a presheaf over the field* k *(in either the flat or étale topologies), and assume that* F *commutes with disjoint sums of schemes. (Every sheaf has this property!) If* K/k *is a normal, separable (but, not necessarily finite) extension with Galois group* G, *then* $F(K)$ *is a* G-*module in a natural way and there exists a canonical isomorphism*

$$H^r_{e \text{ or } f}(K/k, F) \xrightarrow{\sim} H^r(G, F(K))$$

for each $r \geqq 0$.

Proof: First assume K/k is a finite extension. We have a natural isomorphism

$$\otimes_k^{n+1} K \longrightarrow \prod_{G^n} K \ ,$$

where G^n denotes the cartesian product of G with itself n-times, and where the algebra on the right is the product of copies of K indexed by the set G^n. This isomorphism is given by

$$a_0 \otimes \cdots \otimes a_n \rightarrow (a_0(\sigma_1 a_1)(\sigma_1 \sigma_2 a_2) \cdots (\sigma_1 \sigma_2 \cdots \sigma_n a_n))$$

where the element on the right is that element of the product occuring in the $<\sigma_1, \ldots, \sigma_n>^{th}$ place. (The proof is just the normal basis theorem.) If $\sigma \in G$, then σ is a k-algebra homomorphism $K \rightarrow K$, so $F(\sigma): F(K) \rightarrow F(K)$ is defined, and this gives $F(K)$ a canonical G-module structure. According to the hypothesis, $F(\prod_{G^n} K) = \prod_{G^n} F(K) = C^n(G, F(K))$; so we obtain a cochain isomorphism $C^n(K/k, F) \xrightarrow{\sim} C^n(G, F(K))$. One verifies easily that this isomorphism commutes with the respective coboundary maps. Thus our Proposition holds for finite extensions. The general case is obtained by passing to the limit. Q.E.D.

PROPOSITION 71. *Let* k *be a field, then the category of sheaves in the étale topology over* k *is equivalent to the category of (continuous)* G_k-*modules. The equivalence is given by* $F \leadsto F(k_s)$, *and it yields an isomorphism of cohomology groups*

$$\begin{aligned} H^r_e(k, F) &\xrightarrow{\sim} H^r(G(k_s/k), F(k_s)) \\ &= H^r(k, F(k_s)) \end{aligned}$$

for all $r \geqq 0$. *In addition, the étale cohomology may be computed by the* *Čech method for all* $r \geqq 0$.

Proof: To prove this, all we need to show is that the category of étale algebras over k is dual to the category of finite G-sets where $G = G_k = G(k_s/k)$. For then, we may apply Theorem 41 to complete the proof. Each finite G-set is the disjoint union of its "connected components"—these being the orbits of elements under G. Each such orbit has the form G/U where U is the stabilizer of the orbit and is an open subgroup of G. Each *étale* algebra is the product of separable field extensions over k. We will use these "dual" decompositions to establish the duality of categories. To do this, we need only associate field extensions with connected G-sets. But if L is a field extension, $L \otimes_k k_s$ is the product of $[L:k]$ copies of k_s and G acts by permuting the factors. We get a transitive (hence, connected) G-set whose stabilizer is $G(k_s/L) = U$; so L corresponds to $G/G(k_s/L)$. This is precisely the Galois Theory, whose duality properties are well-known; so we are done. Q.E.D.

We have now shown that Galois cohomology (and all its applications of the previous chapters) is precisely *étale* cohomology over a field. We also know that on finite separable layers, for sheaves in the *étale* topology, the cohomology groups $H_e^r(K/k, F)$ and $H_f^r(K/k, F)$ agree, and that the Čech method computes the *étale* cohomology (over fields) for all dimensions. It is not true that the Čech method will work in all dimensions for the flat topology, however, enough is true so that use of the Čech method yields substantial information.

THEOREM 42. *Let* F *be a sheaf (in the flat topology) over the field* k *which is representable by a group scheme of finite type over* k. *Then the natural homomorphisms*

$$\check{H}_f^r(k, F) = H^r(\bar{k}/k, F) \longrightarrow H_f^r(k, F)$$

are isomorphisms for all $r \geqq 0$.

Proof: We shall show that both the Čech functors $F \rightsquigarrow \{\check{H}^r(k,F)\}_r$ and the derived functors $F \rightsquigarrow \{H^r(k,F)\}_r$ form universal δ-functors on the *abelian category of commutative group schemes of finite type over* k. (See [GQ] for a proof that this category is actually abelian.) We know that these two supposed δ-functors agree in dimension zero, so this will complete the proof.

Actually, the hardest part of the proof is the part proving that the Čech functors form a δ-functor; so we shall start with this part. Obviously, it will suffice to prove the statement: *If*

$$0 \to F' \to F \to F'' \to 0$$

is an exact sequence of group schemes of finite type over k, *then the sequence*

$$(*) \qquad 0 \to C^r(\bar{k}/k, F') \to C^r(\bar{k}/k, F) \to C^r(\bar{k}/k, F'') \to 0$$

is exact (for all $r \geqq 0$*).*

In any case, the sequence (*) is left exact because it is merely the sequence of points of these group schemes with values in $\otimes_k^{r+1}\bar{k}$. So only the surjectivity on the right need be proved.

If k has characteristic zero, this is trivial in view of Proposition 71 and the cohomology theory of profinite groups. Therefore, we may and do assume that $\operatorname{ch} k = p > 0$. If A denotes the ring $\otimes_k^{r+1} K$ for some finite, normal extension K/k, we may consider the flat topology over Spec A. Since A is flat over k, we obtain the exact sequence of group schemes over A

$$0 \to F' \otimes_k A \to F \otimes_k A \to F'' \otimes_k A \to 0$$

(here $F \otimes_k A$ means the functor represented by $G_F \times_{\operatorname{Spec} k} \operatorname{Spec} A$, where G_F represents F over k). Taking flat cohomology over A, we get the exact sequence

$$(**) \qquad 0 \to F'(A) \to F(A) \to F''(A) \xrightarrow{\ \delta\ } H^1(A, F' \otimes A) .$$

Now for any $\xi \in C^r(\bar{k}/k, F'')$, there is a finite, normal extension K/k and an element $\xi' \in F''(\otimes_k^{r+1} K)$, such that ξ' maps onto ξ under the map induced by the inclusion of algebras $A = \otimes_k^{r+1} K \to \otimes_k^{r+1} \bar{k}$. The exact sequence (**) then shows that ξ' is mapped to some element $\eta' = \delta(\xi')$ in $H^1(A, F' \otimes A)$. We shall show that there exists a field extension L finite and normal over k, containing K, such that the image of η' in $H^1(B, F' \otimes B)$ (where $B = \otimes_k^{r+1} L$) is zero. This will prove the surjectivity on the right in the sequence (*).

For any group scheme of finite type over k, say F', there exists an extension, k', finite over k, such that over k', $F' \otimes_k k'$ has a composition series all of whose factors are either μ_p, α_p, or a smooth group scheme over k'. (See [GDm]). Here, μ_p is the kernel of the map $G_m \xrightarrow{p} G_m$, where $G_m(A) = A^*$—the units of A. So, $\mu_p(A) = \{x \in A \mid x^p = 1\}$—it is the pth roots of unity group scheme. The group scheme α_p is the kernel of the pth power map on G_a, where $G_a(A) = A^+$—the additive group of A. So, $\alpha_p(A) = \{x \in A \mid x^p = 0\}$—the pth roots of zero group scheme. We may obviously assume that our field K of the previous paragraph contains k'. In this case, we can use the half exactness of $H^1(-, F' \otimes -)$ and induction on the length of a composition series for F' to reduce the question to three cases: (a) $F' = \mu_p$, (b) $F' = \alpha_p$, (c) F' is smooth over A.(†)

Case (a). $F' \otimes A = \mu_p$ over A, so there is an exact sequence (over A)

$$0 \to F' \otimes A \to G_m \xrightarrow{p} G_m \to 0.$$

Compute the cohomology over A, and obtain

(†)Recall this argument: Say $0 \to F_0 \to F' \to F_1 \to 0$ is exact and F_0, F_1 have the required property. If $\eta' \in H^1(A, F' \otimes A)$, then there is a B over A such that $\bar{\eta}' \in H^1(A, F_1 \otimes A)$ goes to zero in $H^1(B, F_1 \otimes B)$. If $\tilde{\eta} =$ image of η' in $H^1(B, F' \otimes B)$, then it follows that $\tilde{\eta}$ comes from some η_0 in $H^1(B, F_0 \otimes B)$. But then there is a C such that η_0 goes to zero in $H^1(C, F_0 \otimes C)$, and one easily sees that the original element η' goes to zero in $H^1(C, F' \otimes C)$.

$$H^0(A, G_m) \xrightarrow{p} H^0(A, G_m) \longrightarrow H^1(A, F' \otimes A) \longrightarrow H^1(A, G_m) \ .$$

Now $H^1(A, G_m) = \check{H}^1(A, G_m)$, and it is known from descent theory [GG] that $\check{H}^1(A, G_m)$ classifies the principal homogeneous spaces for G_m over A. Because $A = \otimes_k^{r+1} K$, A is a product of Artin Local rings; so $H^1(A, G_m)$ is the product of $H^1(A_j, G_m \otimes A_j)$ where the A_j are Artin Local rings. By descent theory, the groups $H^1(A_j, G_m \otimes A_j)$ vanish (see [GG])—this is the general form of Hilbert's Theorem 90—and, hence $H^1(A, G_m) = (0)$. Our cohomology sequence yields the exact sequence

$$(\otimes_k^{r+1} K)^* \xrightarrow{p} (\otimes_k^{r+1} K)^* \longrightarrow H^1(A, F' \otimes A) \longrightarrow 0 \ .$$

Let $\xi_0 = \Sigma_i\, a_{i0} \otimes \cdots \otimes a_{ir}$ be the pre-image of η' in $(\otimes_k^{r+1} K)^*$, then if L is the finite extension of K obtained by adjunction of the pth roots of $a_{i0}, \ldots, a_{ir}$ for all i, we see that ξ_0 becomes a pth power in $(\otimes_k^{r+1} L)^* = B^*$, so that η' goes to zero in $H^1(B, F' \otimes B)$.

Case (b). $F' \otimes A = \alpha_p$, over A; so there is an exact sequence (over A)

$$0 \longrightarrow F' \otimes A \longrightarrow G_a \xrightarrow{p} G_a \longrightarrow 0 \ .$$

Now we repeat word for word the argument of Case (a) (using the fact that Hilbert Theorem 90 is valid for G_a, as well [GG]), and obtain the proof of Case (b).

Case (c). $F' \otimes A$ is smooth over A. As before, $H^1(A, F' \otimes A)$ classifies the principal homogeneous spaces for $F' \otimes A$ over A. The element η' determines such a space, say M; and there is a finite faithfully flat A-algebra Λ which *splits* M, [GG], i.e., for which $M \otimes_A \Lambda \xrightarrow{\sim} F' \otimes_k \Lambda = (F' \otimes_k A) \otimes_A \Lambda$. We get the following diagram

$$\begin{array}{ccccc}
M \quad F' \times_k \operatorname{Spec} A & \longleftarrow & F' \times_A \operatorname{Spec} \Lambda & \xrightarrow{\sim} & M \times_A \operatorname{Spec} \Lambda \\
\theta \searrow \quad \downarrow \zeta & & \downarrow \zeta \times 1 = \theta \times 1 & & \\
\operatorname{Spec} A & \xleftarrow{\pi} & \operatorname{Spec} \Lambda & &
\end{array}$$

where the morphism π is a finite, faithfully flat morphism, and ζ, $\zeta \times 1$ are smooth morphisms. It follows that $\theta \times 1$ is a smooth morphism and, as π is faithfully flat (and finite), the descent theory [GD_2] shows that θ *is smooth* as well.

Now A is a direct product of its local components A_j, and the residue fields of all the A_j are exactly K. The reductions of M over each residue field (call these reductions M_j) obtain rational points in some finite extensions of the residue fields in question. Let L be large enough to contain all these (finitely many) field extensions, then each reduction of M obtains a section over L (i.e., a rational point); hence *as* M *is smooth,* M obtains a section over each B_j. (B_j is the jth local component of $\otimes_k^{r+1} L$.) The product of these sections is one for M over B, which means that η' goes to zero in $H^1(B, F' \otimes B)$.

We now know that the Čech sequence is a δ-functor, and of course the derived sequence is automatically a δ-functor. To prove the universality, first note that for any sheaf F, any $r > 0$, and any $\xi \in H^r(k, F)$, there exists a finite extension K/k such that ξ goes to zero under the map

$$\text{res}: H^r(k, F) \longrightarrow H^r(K, F \otimes K)$$

(this is the restriction map in the *étale* case). To see this, observe that for Čech cohomology it is trivial; and so in dimension 1 it is valid for the derived functors. Now dimension shift by embedding F in an injective sheaf Q in the obvious way to obtain the statement for all $r \geqq 1$.

If K/k is a given finite extension, and if F is a presheaf over K, we define $\pi_*(K \to k)(F)$ by

$$\pi_*(K \to k)(F)(A) = F(A \otimes_k K)$$

and we obtain a presheaf over k. (In the *étale* case, this is precisely the induced module π_* from the subgroup $H = G(k_s/K)$ to the big group $G = G(k_s/k)$.) We call $\pi_*(K \to k) F$ the *direct image* of F. If F is the sheaf corresponding to a group scheme of finite type over K; then $\pi_*(K \to k)F$ is representable [GH] by a group scheme of finite type over k. Moreover, there is the canonical injection

$$F \longrightarrow \pi_*(K \to k)(F \otimes K)$$

(because $F \otimes K$ is really $\pi^*(K \to k)\,F$) for all sheaves F over k.

If ξ belongs to $\check{H}^r(k, F)$, and if ξ comes from $H^r(K/k, F)$ under "inflation," then one verifies by a simple, explicit computation that ξ goes to zero under the map

$$H^r(K/k, F) \to H^r(K/k, \pi_*(K \to k)(F \otimes K)) \ .$$

One sees from this that the Čech functors (for $r > 0$) are "locally effacable" in the sense of [B], and by that paper they are universal.

For the derived functors, one finds from the representability of $\pi_*(K \to k)\,F$ *and* the fact that k is a *field*, that π_* is an *exact functor from sheaves over* K *to sheaves over* k. The argument of Chapter II, §4 concerning spectral sequences now shows that the *Leray spectral sequence*

$$H^p(k, R^q\pi_*(K \to k)\,F) \implies H^*(K, F)$$

(just the usual spectral sequence for composed functors!) degenerates and yields *Shapiro's Lemma: The map*

$$\text{sh}\colon\ H^p(k, \pi_*(K \to k)\,F \otimes K) \longrightarrow H^p(K, F)$$

is an isomorphism for all $p \geqq 0$. But the diagram

$$\begin{array}{ccc} H^r(k, F) & \xrightarrow{\text{res}} & H^r(K, F) \\ \Big\uparrow \text{id} & & \Big\uparrow \text{sh} \\ H^r(k, F) & \longrightarrow & H^r(k, \pi_*(K \to k)(F \otimes K)) \end{array} \ ,$$

and the fact that res kills elements ξ for sufficiently large K, show that $H^r(k, -)$ is locally effacable for $r > 0$. Once again Buchsbaum's result yields the universality, and the proof is complete. Q.E.D.

COROLLARY. *The flat cohomology groups* $H^r(k, G_m)$ *may be computed as the cohomology of the complex*

$$0 \to \bar{k}^* \to (\bar{k} \otimes \bar{k})^* \to (\bar{k} \otimes \bar{k} \otimes \bar{k})^* \to \cdots$$

(the so-called Amitsur complex [RZ]*). In addition, the flat cohomology groups* $H^r(k, G_a)$ *vanish for positive* r.

Proof: The second statement follows because the vanishing is known for the Cech cohomology of G_a, [GG].

There is a connection between *étale* and flat cohomology which is very useful to know. The most general theorem of this type is due to Grothendieck [GB], however, in the case of cohomology over a ground *field* substantial simplifications occur and we follow the presentation given in [SSS].

PROPOSITION 72. *Let* k *be a field, and let*

$$0 \to G' \to G \to G'' \to 0$$

be an exact sequence of commutative group schemes of finite type over k. *Assume that* G' *is smooth over* k. *Then* G *is smooth over* k *if and only if* G'' *is smooth over* k.

Proof: If k_s denotes the separable closure of k, then the sequence

$$0 \to G' \otimes k_s \to G \otimes k_s \to G'' \otimes k_s \to 0$$

is exact. Since k_s is faithfully flat over k, smoothness over k_s is equivalent to smoothness over k. Therefore, we may assume at the outset that k is separably closed.

Now in order to prove that a given group scheme G of finite type is smooth over k, we need to prove that the map $G(A) \to G(A/\mathfrak{A})$ is always surjective, where A is a local Artin ring over k and $\mathfrak{A}$ is an ideal ($\neq (1)$) of A. One sees this by using the fact that G possesses a composition series (over $\bar{k}$) whose factors are smooth, or μ_p, or α_p, and by using the universal mapping criterion for smoothness [GD_2].

So given the pair A, $A/\mathfrak{A}$ as above, we obtain the commutative diagram with exact rows

$$\begin{array}{ccccccccc} 0 & \longrightarrow & G'(A) & \longrightarrow & G(A) & \longrightarrow & G''(A) & \longrightarrow & H^1(A, G'\otimes A) \\ & & \downarrow{\scriptstyle \phi_1} & & \downarrow{\scriptstyle \phi_2} & & \downarrow{\scriptstyle \phi_3} & & \downarrow{\scriptstyle \theta} \\ 0 & \longrightarrow & G'(A/\mathfrak{A}) & \longrightarrow & G(A/\mathfrak{A}) & \longrightarrow & G''(A/\mathfrak{A}) & \longrightarrow & H^1(A/\mathfrak{A}, G'\otimes A) \end{array} .$$

The right hand groups classify principal homogeneous spaces, as is usual. However, any such space has a trivial reduction because it is smooth and the residue field of A (hence of $A/\mathfrak{A}$) contains k which is separably closed. By smoothness, we may lift the rational points of these reductions back to A (resp. $A/\mathfrak{A}$). This proves each P.H.S. over A (resp. $A/\mathfrak{A}$) is trivial; so the right hand groups vanish. Now diagram chasing shows that ϕ_2 is surjective if and only if ϕ_3 is surjective. Q.E.D.

THEOREM 43. *Let* k *be a field and let* G *be a commutative group scheme of finite type over* k. *If* G *is smooth over* k, *then the edge homomorphisms*

$$e_r : H^r_e(k, G) \to H^r_f(k, G)$$

are isomorphisms for every $r \geq 0$.

Proof: We take the limit of spectral sequences (SS1) over all separable extensions K/k and obtain the Hochschild-Serre spectral sequence

$$H^p_f(k_s/k, \mathcal{H}^q(G)) \Longrightarrow H^*_f(k, G) .$$

By Proposition 70, we deduce that our sequence is

$$H^p(k_s/k, H^q_f(k_s, G)) \Longrightarrow H^*_f(k, G)$$

where $H^p(k_s/k, -)$ means cohomology in the Galois sense. Therefore, Theorem 43 will follow from the statement: *If* G *is smooth over* k, *and if* k *is separably closed, then* $H^r_f(k, G)$ *vanishes for positive* r.

The latter statement is already true for $r = 1$ according to the proof of Proposition 72. Assume the statement for $r-1$ and all smooth G of finite type over k. If $\xi \in H^r(k, G)$ is given, we choose a finite field extension K/k so that $\mathrm{res}(\xi)$ vanishes in $H^r(K, G \otimes K)$. Let G_* be the group scheme $\pi_*(K \to k)(G \otimes K)$. One knows that G_* is smooth over k, [GD_2], so the cokernel C in the sequence

$$0 \to G \to G_* \to C \to 0$$

is also smooth over k by Proposition 72. However, cohomology yields the commutative diagram

$$\begin{array}{ccccc} (0) = H_f^{r-1}(k, C) & \longrightarrow & H_f^r(k, G) & \longrightarrow & H_f^r(k, G_*) \\ & & & \overset{\text{res}}{\searrow} & \downarrow \text{sh} \\ & & & & H_f^r(K, G \otimes K) \end{array},$$

and, since sh is an isomorphism, the induction hypothesis shows that $\xi \in H^{r-1}(k, C) = (0)$. Q.E.D.

COROLLARY 1. *If* k *is a local field, then* $H_f^2(k, G_m)$ *is isomorphic via the invariant map to* $\mathbf{Q}/\mathbf{Z}$.

Proof: The group scheme G_m is smooth over k, so we may apply Theorems 37 and 43. Q.E.D.

From now on we shall omit the subscript "f" on cohomology groups —all our groups shall be taken in the flat topology. We know that the Galois cohomology is subsumed in this, and that the values computed by the latter cohomology agree with the results in the flat topology.

To finish this section and prepare for the next, we shall now discuss cup-products. As is usual in these matters, there is a functorial approach, (see, for example, [HJ]) but because of space limitations we shall deal with explicit formulas. This is possible for group schemes in view of Theorem 42.

If F, F′ are two sheaves over k, then *a pairing* of F, F′ into a third sheaf F″ is a morphism $F \times F' \to F''$ such that for each T, the induced map $F(T) \times F'(T) \to F''(T)$ is a pairing of ordinary groups. If M is a field extension of k (algebraic, but not necessarily finite), let $T_r(M)$ be the Spec of the k-algebra $\otimes_k^{r+1} M$. There is a morphism from $T_{r+s}(M)$ to the product $T_r(M) \times T_s(M)$ given by the map of rings

$$(\alpha_0 \otimes \cdots \otimes \alpha_r) \otimes (\beta_0 \otimes \cdots \otimes \beta_s) \longmapsto \alpha_0 \otimes \cdots \otimes \alpha_{r-1} \otimes \alpha_r \beta_0 \otimes \beta_1 \otimes \cdots \otimes \beta_s$$

Call this morphism θ_{rs}. Let F, F', F'' be commutative group schemes of finite type over k, and let P be a pairing of F and F' into F''. If $f \in C^r(M/k, F)$, $g \in C^s(M/k, F')$, we set

$$f \cup_P g = P \circ f \otimes g \circ \theta_{rs} .$$

This yields a mapping

$$C^r(M/k, F) \times C^s(M/k, F') \to C^{r+s}(M/k, F'') ,$$

and an easy computation shows that the usual formula

$$\delta(f \cup_P g) = \delta f \cup_P g + (-1)^r (f \cup_P \delta g)$$

holds. Therefore, upon passing to cohomology, we obtain our cup product.

When M is chosen to be k_s, then the isomorphism of Proposition 71 takes the above cup product into the usual one for Galois theory. Moreover, the axioms for a cup-product are satisfied by this explicit cup-product; hence it is the unique product on cohomology induced by the pairing $F \times F' \to F''$. From these remarks one sees that all the formulas connecting transfer, restriction, and inflation with cup products are valid here, as well.

§4. Finite Group Schemes and their cohomology over a local field

All group schemes will be commutative—so we shall omit any reference to this property. A finite group scheme over k is a group scheme G which, as scheme, is finite over k. Thus, G = Spec A, for some finite dimensional k-algebra, A. The group multiplication, inverse, and identity make A into a Hopf algebra over k, *so the linear dual of* A, say A^D, *is also a Hopf algebra. It follows that* Spec $A^D = G^D$ *is a finite group scheme, called the Cartier Dual of* G, *of the same rank as* G *over* k.

Cartier Duality is extremely important; it will play the same role for the coming duality as the "upper (d)" duality played in the theory of Poincaré groups. Now if F', F'' are sheaves, then one can make the *sheaf of homomorphisms from* F' *to* F''. This sheaf, $\mathbf{Hom}(F', F'')$, is defined by

$$\mathbf{Hom}(F', F'')(T) = \mathrm{Hom}_T(F' \times T, F'' \times T) \ .$$

If $F \times F' \to F''$ is a pairing of sheaves, there is an induced pairing $(F \times T) \underset{T}{\times} (F' \times T') \to F'' \times T$; hence, we obtain a mapping

$$F(T) = (F \times T)(T) \to \mathrm{Hom}_T(F' \times T, F'' \times T)$$

for each T. Thus, a pairing $F \times F' \to F''$ induces a homomorphism of sheaves

$$F \to \mathbf{Hom}(F', F'') \ .$$

If G is a finite group scheme over the field k, and if G^D is its Cartier dual, *then there is a canonical pairing* $G \times G^D \to \mathbf{G}_m$. Indeed, if $A = \Gamma(G, \mathcal{O}_G)$, then such a pairing is a special sort of mapping from $k[X, Y]/(XY - 1)$ to $A \otimes A^D$, say P. Such a map may be given by $P(X) = \Sigma_j \, \xi_j \otimes \xi_j^D$, where $\{\xi_j\}$ ranges over a k-base for A and $\{\xi_j^D\}$ is the dual base. We therefore obtain *a canonical homomorphism*

$$\gamma_G : G^D \to \mathbf{Hom}(G, \mathbf{G}_m) \ ,$$

and one may easily verify (using the fact that k is a field) *that* γ_G *is an isomorphism.* This is true more generally, and the reader may examine [O] for generalizations.

From its first definition, Cartier duality is an involutory functor which is also exact. If we use this fact, we may prove

PROPOSITION 73. *Let* G *be a finite group scheme over the field* k. *Then* G *possesses a unique composition series over* k

$$G \supseteq G_0 \supseteq G_m \supseteq (0)$$

for which

(a) *The quotient* G/G_0 *is étale*

(b) *The quotient* G_0/G_m *is of additive type*

(c) *The quotient* $G_m = G_m/(0)$ *is of multiplicative type*

(Here, "*étale*" means *étale* as a scheme, "additive type" means connected together with its dual, and multiplicative type means the dual is *étale*.)

Proof: Let G_0 be the connected component of identity in G, then one knows [O, Ga] that G/G_0 is *étale*, call it G_s. Consider G_0^D, it may not be connected; so the above decomposition may be applied and it yields the exact sequence

$$0 \to (G_0^D)_0 \to G_0^D \to (G_0^D)_s \to 0 \ .$$

Dualize this sequence to obtain

$$0 \to ((G_0^D)_s)^D \to G_0 \to ((G_0^D)_0)^D \to 0 \ .$$

Let $G_m = ((G_0^D)_s)^D$, $G_a = ((G_0^D)_0)^D$. Then G_m^D is *étale*, so G_m is a finite group scheme of multiplicative type. All we need show is that G_a is of additive type. However, $G_a^D = (G_0^D)_0$ is connected; and as G_a is a homomorphic image of the connected scheme G_0 it is connected itself. Q.E.D.

Examples of the various types of finite group schemes are given by:

(a) *étale* G. Let M be a $G(k_s/k)$-module, finite as an abelian group. Let $\mathfrak{A}(M)$ be the k-algebra of all functions on M to k_s, say f, such that $f(\sigma m) = \sigma f(m)$ for all $\sigma \in G(k_s/k)$, all $m \in M$. Then $G = \mathbf{M} = \operatorname{Spec} \mathfrak{A}(M)$ is *étale*, and *all étale finite group schemes arise this way.* (For the proof look closely at Proposition 71.)

(b) *multiplicative* G. Let G_m be the multiplicative group scheme, let n be an integer. The kernel of the nth power map of G_m to itself is a finite group scheme μ_n, of multiplicative type, *the* n^{th} *roots of unity scheme.* (Observe: $\mu_n(A) = \{x \in A \mid x^n = 1\}$). Second example: M^D for M as in (a).

(c) *additive* G. The group scheme α_p is additive, moreover α_p is self-dual.

If k is a local field, the cohomology of a finite group scheme G over k mirrors the topological properties of k. Proposition 73 shows *that if* k *is* ANY *field,* $H^0(k, G)$ *is finite when* G *is a finite group scheme over* k. If k is local, let $\bar{k}$ be its algebraic closure, $\mathcal{O}$ its ring of integers, and $\mathcal{O}_{\bar{k}}$ the ring of integers in $\bar{k}$. In $\otimes_k^n \bar{k}$, define $\mathbf{O}_n$ by:

$$\mathbf{O}_n = \{\alpha \in \otimes_k^n \bar{k} \mid \alpha = \sum_i \xi_1^{(i)} \otimes \cdots \otimes \xi_n^{(i)}, \text{ with } \xi_j^{(i)} \in \mathcal{O}_{\bar{k}} \text{ for all i, all j}\} .$$

Clearly, $\mathbf{O}_n$ is an $\mathcal{O}$-subalgebra of $\otimes_k^n \bar{k}$. If $m \in \mathbf{Z}$, we set

$$\mathbf{O}_n^{(m)} = t^m \mathbf{O}_n, \quad \text{t a given uniformizer in k} .$$

Then $\otimes_k^n \bar{k} = \bigcup_{m=-\infty}^{\infty} \mathbf{O}_n^{(m)}$; and, given $\xi \in \otimes_k^n \bar{k}$, define $|\xi|$ by the equations

$$|\xi| = q^{-z}, \text{ if } \xi \in \mathbf{O}_n^{(z)} \text{ but } \xi \notin \mathbf{O}_n^{(z+1)} .$$

Here, q is the cardinal of k_0—the residue field of k. The following rules are easily established: $|\xi| = 0$ iff $\xi = 0$, $|\xi + \eta| \leqq |\xi| + |\eta|$, $|\alpha\xi| = \|\alpha\|_k |\xi|$, for all $\alpha \in k$, $|\xi\eta| \leqq |\xi|\,|\eta|$. *In this way,* $\otimes_k^n \bar{k}$ *becomes a normed algebra over* k.

If G is a group scheme of finite type over k, then G is quasi-projective over k [GDm], so $G(\otimes_k^n \bar{k})$ is naturally a topological group. (Its topology is induced by its embedding in projective space over $\otimes_k^n \bar{k}$. A change of embedding induces a homeomorphism of $G(\otimes_k^n \bar{k})$ with itself.) Therefore, each group $C^n(\bar{k}/k, G)$ (for $n \geqq 0$) is a topological group. The coboundary mapping is given by polynomials so it is continuous, any homomorphism

$G' \to G$ of group schemes yields a continuous homomorphism of the corresponding cochain groups, and one checks that if $G' \to G$ is an injection of group schemes then $C^n(\bar{k}/k, G') \to C^n(\bar{k}/k, G)$ is a homeomorphism into with closed image.

When G is a finite group scheme, $C^0(\bar{k}/k, G)$ is finite (by Proposition 73); hence compact. Therefore, $B^1(\bar{k}/k, G)$ is compact and closed in the closed subgroup $Z^1(\bar{k}/k, G)$ of $C^1(\bar{k}/k, G)$. Now Theorem 42 shows *that the group* $H^1(k, G)$ *has a natural topology for each finite group scheme* G. Also, the mapping $Z^1(\bar{k}/k, G) \to H^1(k, G)$ is closed and open because $B^1(\bar{k}/k, G)$ is compact.

PROPOSITION 74. *If* k *is a local field and if* G *is a finite group scheme over* k, *then for every finite extension* K/k:

(a) $\inf: H^1(K/k, G) \to H^1(k, G)$ *is a homeomorphism into with closed image, and*

(b) $\operatorname{res}: H^1(k, G) \to H^1(K, G \otimes K)$ *is continuous.*

Proof: Inflation is injective and continuous by our remarks above. Let S be a closed subset of $H^1(K/k, G)$, and let S' be its inverse image in $Z^1(K/k, G)$. Then S' is closed in $C^1(K/k, G)$, and, as $K \otimes_k K \to \bar{k} \otimes_k \bar{k}$ is a homeomorphism with closed image, we deduce that the image of S', say S'', in $C^1(\bar{k}/k, G)$ is closed there. Because $B^1(\bar{k}/k, G)$ is compact, the projection of S'' into $H^1(k, G)$ is closed; and this set is precisely $\inf(S)$.

The proof of (b) is entirely similar; it makes use of the fact that the mapping $\bar{k} \otimes_k \bar{k} \to \bar{k} \otimes_K \bar{k}$ is continuous because K/k is *finite*. Q.E.D.

The connected finite group schemes have cohomology groups with nontrivial topological properties. To study these we make use of the *Frobenius morphism.*

If X is a scheme over the field k of characteristic $p > 0$, then raising to the pth power in each stalk of $\mathcal{O}_X$ yields a morphism $\phi_p(X): X \to X$ which is set-theoretically the identity. The commutative diagram

$$\begin{array}{ccc} X & \xrightarrow{\phi_p(X)} & X \\ \downarrow{\scriptstyle \pi_X} & & \downarrow{\scriptstyle \pi_X} \\ \text{Spec } k & \xrightarrow{\phi_p(k)} & \text{Spec } k \end{array}$$

is NOT cartesian, so we are lead to introduce the fibred product

$$< X \times \text{Spec } k, \pi_X, \phi_p > = F_k(X) \ .$$

This is the *Frobenius Scheme of* X over k, it is a scheme over k by its second projection. The mappings $\phi_p(X)$, π_X of X into the factors of $F_k(X)$ yield a functorial map

$$f_{X/k}: X \to F_k(X)$$

of schemes over k called the *Frobenius morphism.* The Frobenius operation is a base extension, so it commutes with products and inverse limits. Moreover, $F_k(X)$ is a group scheme if X is such, and $f_{X/k}$ is then a homomorphism of group schemes. If we repeat the procedure we get the schemes $F_k^n(X)$ defined by

$$F_k^n(X) = F_k(F_k^{n-1}(X)), \ F_k^0(X) = X \ ,$$

and the functorial morphisms $f_{X/k}^{(n)} \ X \to F_k^n(X)$.

PROPOSITION 75. *Let* G *be a finite, connected group scheme over the field* k. *Then the kernel of* $f_{G/k}$ *is never zero unless* G *is trivial. In particular, there exists an integer* $n < \infty$, *such that* $f_{G/k}^{(n)}$ *is the zero map.*

Proof: If $A(G)$ is the affine ring of G, then the affine ring of $F_k(G)$ is $A(G) \otimes_k k$, where k is embedded in itself *via* the pth power. The Frobenius morphism is precisely the ring map given by $a \otimes \lambda \longmapsto a^p \cdot \lambda$, where $a \in A(G)$, $\lambda \in k$. Now let Λ be the ring of "dual numbers" over k, i.e., $\Lambda = k[t]/(t^2)$; then $G(\Lambda)$ is naturally isomorphic to the Zariski tangent space of G at the origin. That is, $G(\Lambda)$ is the vector space dual of $\mathfrak{m}/\mathfrak{m}^2$, where $\mathfrak{m}$ is the maximal ideal of $A(G)$.

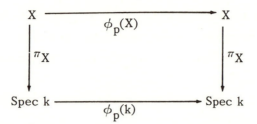

The Frobenius morphism yields a homomorphism $G(\Lambda) \to (F_k G)(\Lambda)$; and, by the above description of $G(\Lambda)$ and $f_{G/k}$, we see that this is always the zero homomorphism. Were $f_{G/k}$ injective, our zero map would have to be injective so that $\mathfrak{m} = \mathfrak{m}^2$. Since $\mathfrak{m}$ is nilpotent, we would deduce $\mathfrak{m} = (0)$. Thus G would be trivial. Q.E.D.

The smallest integer n such that $f^{(n)}_{G/k}$ is the zero map (for G a finite, connected group scheme over k) is called the *Frobenius height* of G, or *infinitesimal rank* of G.

PROPOSITION 76. *Let* G *be a connected finite group scheme of Frobenius height* n. *If* k_n *denotes the field* $k^{p^{-n}}$ $(n > 0)$, *then the map*

$$\inf : H^1(k_n/k, G) \to H^1(k, G)$$

is bijective. In particular, if k *is a local field of characteristic* $p > 0$, *then* $H^1(k, G)$ *is locally compact with the second axiom of countability.*

Proof: From spectral sequence (SS1) we obtain the inflation – restriction sequence

$$0 \to H^1(k_n/k, G) \to H^1(k, G) \xrightarrow{\text{res}} H^1(k_n, G \otimes k_n) .$$

We shall show that res is the zero map, and in view of Proposition 74 this will prove both halves of the current Proposition.

For any $n > 0$, the affine ring of $F^n_k(G)$ is $A(G) \otimes_k k$ where k is embedded in itself by the $p^{n\text{th}}$ power. There is a natural isomorphism $G \otimes k_n \to F^n_k(G)$ given by the map g on their affine rings:

$$g(x \otimes w) = x \otimes w^{p^{-n}}, \quad x \in A(G), \quad w \in k .$$

There is also the natural isomorphism

$$h : \otimes^r_{k_n} \bar{k} \longrightarrow \otimes^r_k \bar{k} ,$$

namely

$$h(\beta_0 \otimes_{k_n} \cdots \otimes_{k_n} \beta_{r-1}) = \beta_0^{p^n} \otimes_k \cdots \otimes_k \beta_{r-1}^{p^n} .$$

Together, g and h yield an *isomorphism*

$$(g, h): C^r(\bar{k}/k_n, G \otimes k_n) \to C^r(\bar{k}/k, F^n_k(G)) \;;$$

hence, we get the *cohomology isomorphism*

$$(g, h)_*: H^r(\bar{k}/k_n, G \otimes k_n) \to H^r(\bar{k}/k, F^n_k(G)) \;.$$

We obtain the diagram

$$\begin{array}{ccc} H^r(k, G) & \xrightarrow{\text{res}} & H^r(k_n, G \otimes k_n) \\ & \searrow^{H^r(f^{(n)}_{G/k})} & \downarrow (g, h)_* \\ & & H^r(k, F^n_k(G)) \end{array}$$

which is clearly commutative. Since the Frobenius height of G is n, $f^{(n)}_{G/k}$ vanishes; hence, res $= 0$ for all $r > 0$. Q.E.D.

There is an equivalence of categories between the abelian, restricted Lie algebras (finite dimensional) over k, and the finite, connected group schemes of Frobenius height one over k. This equivalence is given by associating with each finite group scheme, G, its tangent space $\mathfrak{t}(G) = G(\Lambda)$ —which is an abelian, restricted Lie algebra (recall, *all* our group schemes are commutative); and by associating with each Lie algebra L the Cartier Dual of the Spec of its restricted enveloping algebra U(L). See [Ga] for details. Using this equivalence of categories, we now prove

PROPOSITION 77. *The only simple finite group scheme of additive type over* k *is* α_p.

Proof: A simple group scheme must have a simple dual and it must have Frobenius height one, together with its dual. Therefore, it corresponds to an abelian, restricted Lie algebra which is simple. However, as G is additive, its dual is connected. The Lie algebra $\mathfrak{t}(G)$ is contained in the augmentation ideal of the affine ring of G^D; so, the elements of $\mathfrak{t}(G)$ are nilpotent. By simplicity, they must be nilpotent of index p. This proves that the p-map in $\mathfrak{t}(G)$ is zero.

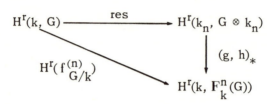

Now, the only simple, Lie algebra with zero p-map is a one-dimensional vector space over k. Thus the corresponding group scheme is α_p. Q.E.D.

COROLLARY. *If* G *is a finite group scheme of additive type over* k, *then* $H^r(k, G) = (0)$ *for* $r \neq 1$.

Proof: The exact sequence of cohomology shows that we may use induction on the length of a composition series for G. By Proposition 77, we are reduced to proving the corollary for the special case $G = \alpha_p$.

The exact sequence $0 \to \alpha_p \to G_a \to G_a \to 0$, yields the cohomology sequence

$$0 \to \alpha_p(k) \to k^+ \to k^+ \to H^1(k, \alpha_p) \to H^1(k, G_a) \to \cdots$$

$$\to \cdots \to H^{r-1}(k, G_a) \to H^r(k, \alpha_p) \to H^r(k, G_a) \to \cdots$$

Since G_a is smooth, its cohomology is computable by *étale* methods; and by Chapter IV, the groups $H^r(k, G_a)$ vanish for $r > 0$. The corollary follows immediately. Q.E.D.

PROPOSITION 78. *Let* G *be an étale finite group scheme over the local field* k. *Then* $H^0(k, G)$ *and* $H^2(k, G)$ *are finite groups, and* $H^r(k, G) = (0)$ *for* $r \geqq 3$. *The group* $H^1(k, G)$ *is discrete in its natural topology. Lastly, if* G *corresponds to a* $G(\bar{k}/k)$*-module with trivial action, then* $C^r(\bar{k}/k, G)$ *is discrete for every* $r \geqq 0$.

Proof: The assertion concerning $H^0(k, G)$ is trivial, and the assertion that $H^r(k, G)$ vanishes for $r \geqq 3$ follows because $\operatorname{cd} k = 2$. (Cf. the tower theorem, Theorem 27, and the remarks in §5 of Chapter IV.) Let us assume the discreteness of $C^r(\bar{k}/k, G)$ for those G with trivial action. If G is an arbitrary *étale* finite group scheme, then there is a $G(\bar{k}/k)$-module corresponding to G (Proposition 71); hence there is a finite field extension K/k such that $G \otimes K$ corresponds to a module with trivial action.

We consider the Hochschild-Serre spectral sequence

$$H^p(K/k, H^q(K, G \otimes K)) \Longrightarrow H^*(k, G) .$$

The sequence of terms of low degree is

$$0 \to H^1(K/k, G) \xrightarrow{\text{inf}} H^1(k, G) \xrightarrow{\text{res}} H^1(K, G \otimes K) \ .$$

By assumption, $H^1(K, G \otimes K)$ is discrete because $G \otimes K$ has trivial action; and $H^1(K/k, G)$ is finite; hence, discrete. But, Proposition 74 implies that $H^1(K/k, G)$ is open in $H^1(k, G)$; and it follows immediately that $H^1(k, G)$ is discrete in its natural topology.

For the finiteness of $H^2(k, G)$, we analyze the spectral sequence more closely. Recall that $H^2(k, G)$ has a composition series whose factors are $E_\infty^{p,q}$ for $p+q=2$, $p, q \geqq 0$; therefore, it suffices to prove $E_\infty^{p,q}$ is finite for these values of p, q. But for $r > \sup(p, q+1)$, we know $E_r^{p,q} = E_\infty^{p,q}$; therefore we deduce $E_2^{2,0} = E_\infty^{2,0}$, $E_3^{1,1} = E_\infty^{1,1}$, $E_4^{0,2} = E_\infty^{0,2}$. Now each $E_r^{p,q}$ is the homology of $E_{r-1}^{p,q}$, thus to prove finiteness it will suffice to show that $E_2^{p,q}$ is finite for $p, q \geqq 0$, $p+q=2$. This is trivial for $E_2^{2,0}$, and for $E_2^{0,2}$ one obtains it as follows: $G \otimes K$ has a composition series all of whose factors are groups Z/qZ for prime numbers q. Since $E_2^{0,2}$ is a subgroup of $H^2(K, G \otimes K)$, it suffices to prove finiteness when $G \otimes K$ is one of the groups Z/qZ. If $q = p = \operatorname{ch} k$, then $H^2(K, G \otimes K) = (0)$ by Corollary 2 of Theorem 22. If $q \neq \operatorname{ch} k$, then we have the exact sequence

$$(*) \qquad 0 \to Z/qZ \to \bar{k}^* \xrightarrow{q} \bar{k}^* \to 0$$

(which amounts to choosing a non-canonical K-isomorphism from Z/qZ to μ_q); and the cohomology sequence yields

$$0 \to H^2(K, Z/qZ) \to \operatorname{Br}(K) \xrightarrow{q} \operatorname{Br}(K) \ .$$

Since $\operatorname{Br}(K) = Q/Z$, the case of $E_2^{0,2}$ is settled.

For the group $E_2^{1,1} = H^1(K/k, H^1(K, G \otimes K))$, the same argument shows that one is reduced to the case of Z/qZ for prime numbers q. If $q \neq \operatorname{ch} k$, then the exact sequence (*) yields the cohomology sequence

$$K^* \xrightarrow{q} K^* \to H^1(K, Z/qZ) \to 0 \ ;$$

and since K is a local field, K^*/K^{*q} is a finite group (Proposition 45). This gives the finiteness for $E_2^{1,1}$ if $q \neq \operatorname{ch} k$. If $q = \operatorname{ch} k = p > 0$, then there is the exact sequence

$$0 \to \mathbf{Z}/p\mathbf{Z} \to \mathbf{G}_a \xrightarrow{\wp} \mathbf{G}_a \to 0$$

where $\wp(x) = x^p - x$. By cohomology, we obtain $H^1(K, \mathbf{Z}/p\mathbf{Z}) = K^+/\wp(K^+)$ —an infinite group. The exact sequence

$$0 \to \wp(K^+) \to K^+ \to K^+/\wp(K^+) \to 0$$

shows that $H^1(K/k, K^+/\wp(K^+)) \hookrightarrow H^2(K/k, \wp(K^+))$. But,

$$0 \to \mathbf{Z}/p\mathbf{Z} \to K^+ \to \wp(K^+) \to 0$$

implies that

$$0 \to H^2(K/k, \wp(K^+)) \to H^3(K/k, \mathbf{Z}/p\mathbf{Z})$$

is exact; so $H^2(K/k, \wp(K^+))$ is finite. In this way, we see that $E_2^{1,1}$ is finite.

There remains only the assertion that the groups $C^r(\bar{k}/k, G)$ are discrete if G has trivial action. Let $A(G)$ be the k-algebra whose Spectrum is G. Because G is trivial as $G(\bar{k}/k)$-module, there are generators $b_1, \ldots, b_n$ of $A(G)$ such that $b_i b_j = \delta_{i,j} b_i$, and $\Sigma_{i=1}^n b_i = 1$. Therefore, the group $C^r(\bar{k}/k, G)$ is precisely the subset of $\otimes^{r+1}\bar{k} \times \cdots \times \otimes^{r+1}\bar{k}$ consisting of those n-tuples $y_1, \ldots, y_n$ such that $y_i y_j = \delta_{ij} y_i$ and $\Sigma\, y_i = 1$. Because $y_i^2 = y_i$, we get

$$|y_i| = |y_i^2| \leqq |y_i|^2 \;;$$

that is,

$$y_i \neq 0 \implies |y_i| \geqq 1 \,.$$

Let U be the set $\{\langle x_1, \ldots, x_n\rangle \mid |x_1| < 2 \text{ and } |x_j| < 1 \text{ for } j > 1\}$, then U is open in $\otimes^{r+1}\bar{k} \times \cdots \times \otimes^{r+1}\bar{k}$, and we see that

$$U \cap C^r(\bar{k}/k, G) = \{\langle 1, 0, \ldots, 0\rangle\} \,.$$

Since $\langle 1, 0, \ldots, 0\rangle$ is the unit element of $C^r(\bar{k}/k, G)$, the cochain groups $C^r(\bar{k}/k, G)$ are discrete, as asserted. Q.E.D.

For the duals of *étale* group schemes, we have

PROPOSITION 79. *Let* G *be a finite group scheme of multiplicative type over the local field* k. *Then* $H^0(k, G)$ *and* $H^2(k, G)$ *are finite groups, and* $H^1(k, G)$ *is a compact group.*

Proof: The statement concerning $H^0(k, G)$ is trivial. For the result on 1-cohomology, observe that G is the direct sum of G_p and G', where G_p is a p-group scheme ($p = \mathrm{ch}\, k$), and G' is prime-to-p. The group scheme G' is *étale* so by Proposition 78, we may ignore it (because its 1-cohomology is finite!). Therefore, assume G is a p-group scheme, then its dual, G^D, is *étale*; and hence, is trivialized in some finite extension K/k. If follows that $G \otimes K$ is K-isomorphic to the product of copies of μ_{p^r} for various integers r; hence

$$H^1(K, G \otimes K) = \prod_{j=1}^{m} H^1(K, \mu_{p^{r_j}}) .$$

We shall show that $H^1(K, \mu_{p^r})$ is topologically isomorphic to K^*/K^{*p^r}, which will show that $H^1(K, G \otimes K)$ is compact.

From Proposition 76, it follows that

$$\mathrm{inf}: H^1(K_r/K, \mu_{p^r}) \to H^1(K, \mu_{p^r})$$

is a topological isomorphism. Now the natural map

$$Z^1(K_r/K, \mu_{p^r}) \to H^1(K_r/K, \mu_{p^r})$$

is bijective because $B^1(K_r/K, \mu_{p^r}) = \mu_{p^r}(K_r) = (0)$. Moreover, the natural injection

$$0 \to Z^1(K_r/K, \mu_{p^r}) \to Z^1(K_r/K, G_m)$$

shows that every cocycle for μ_{p^r} is one for G_m. Therefore, Hilbert Theorem 90 in the layer K_r/K shows immediately that topologically and algebraically

$$Z^1(K_r/K, \mu_{p^r}) \xrightarrow{\sim} \{\xi \otimes \frac{1}{\xi} \mid \xi \in K_r^*\} .$$

But the map $\xi \otimes \frac{1}{\xi} \to \xi K^*$ is an algebraic isomorphism of $Z^1(K_r/K, \mu_{p^r})$ and K_r^*/K^*, and the inverse map is continuous as one sees from the identity

$$\xi \otimes \frac{1}{\xi} - 1 \otimes 1 = 1 \otimes (\frac{1}{\xi} - 1) + (\xi - 1) \otimes 1 + ((\xi - 1) \otimes 1)(1 \otimes (\frac{1}{\xi} - 1)).$$

If we apply the $p^{r\text{th}}$ power map, we obtain the continuous isomorphism

$$K^*/K^{*p^r} \to Z^1(K_r/K, \mu_{p^r}) = H^1(K_r/K, \mu_{p^r}) ;$$

which is actually a homeomorphism because K^*/K^{*p^r} is compact.

In the general case, we have the exact sequence

$$0 \to H^1(K/k, G) \xrightarrow{\text{inf}} H^1(k, G) \xrightarrow{\text{res}} H^1(K, G \otimes K)^{\mathfrak{G}(K/k)} .$$

Since G is connected, and since $\mathfrak{G}(K/k)$ acts continuously on $H^1(K, G \otimes K)$, it follows that $H^1(k, G)$ is compact.

The two cohomology statement is proved as before by analyzing the spectral sequence. Both $E_2^{0,2}$ and $E_2^{2,0}$ are obviously finite (the latter vanishes for p-group schemes!), so the only problem is

$$E_2^{1,1} = H^1(K/k, H^1(K, G \otimes K)) .$$

As above, $G \otimes K$ is a product of μ_{p^r}'s, and we are thereby reduced to that case. We apply cohomology to the exact sequence

$$0 \to K^{*p^r} \to K^* \to K^*/K^{*p^r} \to 0 ,$$

and obtain the monomorphism

$$0 \to H^1(K/k, K^*/K^{*p^r}) \to H^2(K/k, K^{*p^r}) .$$

However, the mapping $K^* \to K^{*p^r}$ is an isomorphism, and this shows that $H^2(K/k, K^{*p^r})$ is finite; so we deduce the finiteness of $H^1(K/k, H^1(G \otimes K))$.

Q.E.D.

We now put together Propositions 76, 78, and 79 and prove

PROPOSITION 80. *Let* G *be a finite group scheme over the local field* k. *Then* $H^0(k, G)$ *and* $H^2(k, G)$ *are finite groups and* $H^1(k, G)$ *is a locally compact group with the second axiom of countability.*

Proof: As usual, the assertion for H^0 is trivial. Consider the exact sequences

$$(*) \qquad 0 \to G_0 \to G \to G_s \to 0$$

$$(**) \qquad 0 \to G_m \to G_0 \to G_a \to 0$$

(Proposition 73), and apply cohomology to the first. We get

$$H^1(k, G_0) \xrightarrow{\theta_1} H^1(k, G) \xrightarrow{\theta_2} H^1(k, G_s) \to H^2(k, G_0) \to H^2(k, G) \to H^2(k, G_s) \ .$$

Now (**) implies $H^2(k, G_m) \to H^2(k, G_0) \to H^2(k, G_a)$ is exact, and Proposition 79 and the corollary to Proposition 77 show that $H^2(k, G_0)$ is finite. It follows from this and Proposition 78 that $H^2(k, G)$ is finite.

The maps θ_1, θ_2 are continuous, and θ_2 is open and closed because $H^1(k, G_s)$ is discrete. It is not hard to see that θ_1 is a closed map, so we deduce that $H^1(k, G)$ is a topological group with a locally compact subgroup and quotient group. It follows that $H^1(k, G)$ is locally compact, and it is clearly second countable. Q.E.D.

§5. Tate-Nakayama Duality

Let k be a field, k_s its separable closure, and $\underline{G}$ the Galois group of k_s/k. If M is a $\underline{G}$-module, the group algebra, $k_s(M)$, of M over k_s is a $\underline{G}$-module in a natural way *via*

$$\sigma\left(\sum_m \lambda_m \cdot m\right) = \sum_m (\sigma\lambda_m) \cdot (\sigma m) \ , \qquad \sigma \in \underline{G} \ .$$

Let k_M be the fixed ring of $k_s(M)$ under this action of $\underline{G}$ and let Spec k_M be denoted X_M —it is a group scheme over k. If M is a finitely generated $\underline{G}$-module, then [SA, Lemma 10, Chapter III] shows that X_M is a noetherian scheme. We shall call X_M a *diagonalizable* group scheme over k whenever M is finitely generated. (The general theory of such group schemes is developed in [GDm].) If the torsion part of M is of rank prime-to-p ($p = \operatorname{ch} k$), then X_M is a smooth group scheme over k; hence, Theorem 43 shows that the *cohomology of* X_M *is naturally isomorphic to the Galois cohomology of the* $\underline{G}$*-module* $X_M(k_s) = \operatorname{Hom}_Z(M, k_s^*)$.

If X_M is a diagonalizable group scheme over k, then the functor $A \rightsquigarrow \operatorname{Hom}_{A\text{-groups}}(X_M \times \operatorname{Spec} A, G_m \times \operatorname{Spec} A)$ is a *sheaf* in the flat topology over k; we denote it by $\mathbf{M}$. (When M is a finite group this notation agrees with previous notation introduced in §4 for *étale* finite group schemes.) Moreover, $\mathbf{M}$ is really an *étale* sheaf in that it is ε^*F for some sheaf F in the *étale* topology over k. (Here, ε is the identity map from Spec k with the flat topology to Spec k with the *étale* topology–obviously a "continuous" mapping.) The verification that $\mathbf{M}$ is an *étale* sheaf is very easy and will be omitted–it also follows from the criterion proved in [SSh].

Now an *étale* sheaf remains *étale* after a base extension; so extending the base to k_s, we see that $\mathbf{M} \otimes k_s$ is *étale*. But then, the Čech groups $H^r(B/A, \mathbf{M} \otimes k_s)$ vanish for $r > 0$ and all finite algebras A, B over k_s such that B covers A. Thus, $\mathbf{M} \otimes k_s$ is a *flasque* sheaf over k_s [AG], so its cohomology vanishes; and the spectral sequence

$$H^p(\underline{G}, H^q(k_s, \mathbf{M} \otimes k_s)) \Longrightarrow H^*(k, \mathbf{M})$$

degenerates. We have proved that *the Galois cohomology of* $\mathbf{M}(k_s) = M$ *and the flat cohomology of* $\mathbf{M}$ *are isomorphic.*

To prove the main theorem of this section, we shall have to know that s.c.d. $k = 2$ for local fields k. For our purposes this is most easily seen by using a theorem of Tate which computes the dualizing module for such fields. (This module exists by Propositions 65 and 78.)

THEOREM 44 (Tate). *If* k *is a local field and* $\underline{G}$ *is its Galois group, then the dualizing module for* $\underline{G}$ *is* $\operatorname{dir\,lim}_n \mu_n$, i.e., *the "union" of all the* μ_n.

Proof: (Serre) Let I be the dualizing module for $\underline{G}$ and write I_n for $\ker(I \xrightarrow{n} I)$. For any open subgroup U of $\underline{G}$, we know that I is its dualizing module, and from this we get

$$\mathrm{Hom}_U(\mu_n, I) = H^0(U, \mu_n^{(d)}) = H^2(U, \mu_n)^d .$$

The exact sequence $0 \to \mu_n \to G_m \xrightarrow{n} G_m \to 0$ and Theorem 37 show that $H^2(U, \mu_n) = Z/nZ$; so we deduce that

$$\mathrm{Hom}_U(\mu_n, I_n) = \mathrm{Hom}_U(\mu_n, I) = Z/nZ .$$

In particular, the group $\mathrm{Hom}_U(\mu_n, I_n)$ is independent of U, and has a canonical element corresponding to the canonical generator of Z/nZ. If $\phi_n \in \mathrm{Hom}(\mu_n, I_n)$ is this generator then because there are precisely n elements in Z/nZ and because ϕ_n corresponds to the generator, we find that ϕ_n is an isomorphism of μ_n and I_n. It is also a $\underline{G}$-isomorphism (take $U = \underline{G}$), and if $n|m$ the maps ϕ_n, ϕ_m are compatible by local class field theory. (Generators of Z/nZ and Z/mZ are made to correspond by the fact that restriction multiplies invariant by degree.) Now pass these maps to the limit over all n under the *Artin ordering*: $n \leqq m \iff n \mid m$. We obtain the $\underline{G}$-isomorphism $\operatorname{dir\,lim}_n \mu_n \cong \operatorname{dir\,lim}_n I_n = I$, the latter equality since I is a torsion module. Q.E.D.

As a consequence, we deduce the

COROLLARY. *If* k *is a local field, then* s.c.d. k = 2.

Proof: By Proposition 39 the set $\mu(K)$ is finite for every finite extension K of k. Since $\mu(K) = I^U$ $(U = \underline{G}(k_s/K))$ by Theorem 44, we may apply Proposition 66 to complete the proof. Q.E.D.

The functors $M \rightsquigarrow X_M$, $M \rightsquigarrow \mathbf{M}$ are both exact and faithful; the former is contravariant, the latter covariant. Using these functors and Theorem 44, we can now prove the Tate-Nakayama Duality Theorem.

THEOREM 45 (Tate-Nakayama Duality). *Let* k *be a local field, let* $\underline{G}$ *be the Galois group of* k_s/k, *and let* M *be a torsion free, finitely generated*

$\underline{G}$-module. Let $H^0(k, M)^\wedge$ denote the completion of $H^0(k, M)$ in the topology of subgroups of finite index, and let $H^0(k, X_M)^\wedge$ be the completion of $H^0(k, X_M)$ in the topology of open subgroups of finite index. (The group $H^0(k, X_M)$ has a natural topology as analytic group over k.) Then the duality pairing

$$M \times X_M \to G_m$$

induces the cup-product pairings

$$\theta_r : H^r(k, M) \times H^{2-r}(k, X_M) \to H^2(k, G_m) = Q/Z$$

and (a) *θ_0 is a duality between the compact group $H^0(k, M)^\wedge$ and the discrete group $H^2(k, X_M)$,*

(b) *the groups $H^1(k, M)$, $H^1(k, X_M)$ are finite and θ_1 is a duality,*

(c) *θ_2 is a duality between the discrete group $H^2(k, M)$ and the compact group $H^0(k, X_M)^\wedge$.*

Proof: When $M = Z$ assertion (a) is a trivial consequence of Hasse's Theorem (Theorem 37), assertion (b) is valid as the groups in question vanish, and assertion (c) is precisely the content of the existence theorem of local class field theory. By passing to direct sums, we get the theorem for the case of trivial $\underline{G}$-action on M. The general case can be deduced from arguments involving Tate's Theorem (Theorem 33) and Nakayama's results on cohomological triviality (Chapter V, §2), see [Lg]. However, a slightly more general approach is *via divissage*, and this is the way we shall proceed.

First we shall prove that the groups $H^1(k, M)$ and $H^1(k, X_M)$ are finite. If U is an open normal subgroup of $\underline{G}$ leaving M fixed, then it is an easy consequence of the normal basis theorem that $X_M \otimes K$ is the product of ℓ copies of G_m and $M \otimes K$ is the sum of ℓ copies of Z. Here, K is the fixed field of U and ℓ is the rank of M as Z-module. It follows immediately from Hilbert's Theorem 90 and the opening remarks of this section that $H^1(K, X_M \otimes K)$ and $H^1(K, M \otimes K)$ vanish. Thus, the Hochschild-Serre spectral sequence yields the isomorphisms

$$H^1(K/k, M) \longrightarrow H^1(k, M)$$

$$H^1(K/k, X_M) \longrightarrow H^1(k, X_M) \ .$$

Let r be the degree of K/k, then we know that $H^1(K/k, M)$ and $H^1(K/k, X_M)$ are killed by multiplication by r; therefore so are the groups $H^1(k, M)$ and $H^1(k, X_M)$. Now we have the exact sequences

$$0 \longrightarrow M \xrightarrow{r} M \longrightarrow M/rM \longrightarrow 0$$

$$0 \longrightarrow G \longrightarrow X_M \xrightarrow{r} rX_M \longrightarrow 0$$

(the rX_M is used because in dealing with *étale* cohomology the sequence $0 \to G \to X_M \xrightarrow{r} X_M \to 0$ is *not* exact if $(r, p) \neq 1$, $p = \mathrm{ch}\, k$) and the corresponding cohomology sequences

$$M(k) \to (M/rM)(k) \to H^1(k, M)_r = H^1(k, M) \to 0$$

$$H^0_e(k, rX_M) \to H^1_e(k, G) \to H^1_e(k, X_M)_r = H^1(k, X_M) \to 0 \ .$$

Here, the subscript r means the kernel of multiplication by r. Since we deal with *étale* cohomology, and since both G and M/rM are *finite* group schemes, our cohomology groups are finite, as asserted.

Let U be an open, normal subgroup of $\underline{G}$, and let $\underline{H}$ denote $\underline{G}/U$. If $M = \mathbf{Z}[\underline{H}]$, then we can prove the theorem for M and X_M. To see this, observe that M is $\pi_*(K \to k)\mathbf{Z}$ where K is the fixed field of U; hence, we may apply Shapiro's lemma and the original case of the theorem discussed above for the field K. Then, since cup-product commutes with the isomorphism of Shapiro's lemma, we deduce the result for the given M over k.

In the general case, we can always write M as a quotient of a free, finitely generated $\mathbf{Z}[\underline{H}]$-module F, where $\underline{H}$ has the same significance as above. Thus the sequence

$$0 \to N \to F \to M \to 0$$

is exact, we have the dual sequence

$$0 \to X_M \to X_F \to X_N \to 0 \ ,$$

and the cohomology groups $H^1(k, F)$, $H^1(k, X_F)$ vanish. Moreover, the duality is true for the middle terms F and X_F by our above remarks. If we let the superscript "d" mean duality in Pontrjagin's sense (which for profinite, second countable groups is merely continuous homomorphisms into $\mathbf{Q}/\mathbf{Z}$, i.e., torsion elements of the group of all homomorphisms into $\mathbf{Q}/\mathbf{Z}$), and if $H^r(X)$ denotes $H^r(k, X)$, then the cohomology sequences become

$$0 \to \hat{H}^0(N) \to \hat{H}^0(F) \to \hat{H}^0(M) \to H^1(N) \to 0 \to H^1(M) \to H^2(N) \to H^2(F) \to H^2(M) \to 0$$

and

$$0 \to \hat{H}^0(X_M) \to \hat{H}^0(X_F) \to \hat{H}^0(X_N) \to H^1(X_M) \to 0 \to H^1(X_N) \to H^2(X_M) \to H^2(X_F) \to H^2(X_N) \to 0 \ .$$

(We have used the Corollary to Theorem 44 twice.) Observe that both sequences split naturally into two parts, and so by applying Pontrjagin duality and the remarks above, we get the commutative diagrams

$$\begin{array}{ccccccccc} 0 \longrightarrow & H^1(M) & \longrightarrow & H^2(N) & \longrightarrow & H^2(F) & \longrightarrow & H^2(M) & \longrightarrow 0 \\ & \downarrow v_1 & & \downarrow v_2 & & \downarrow v_3 & & \downarrow v_4 & \\ 0 \longrightarrow & H^1(X_M)^d & \longrightarrow & \hat{H}^0(X_N)^d & \longrightarrow & \hat{H}^0(X_F)^d & \longrightarrow & \hat{H}^0(X_M)^d & \longrightarrow 0 \end{array}$$

$$\begin{array}{ccccccccc} 0 \longrightarrow & H^1(X_N) & \longrightarrow & H^2(X_M) & \longrightarrow & H^2(X_F) & \longrightarrow & H^2(X_N) & \longrightarrow 0 \\ & \downarrow w_1 & & \downarrow w_2 & & \downarrow w_3 & & \downarrow w_4 & \\ 0 \longrightarrow & H^1(N)^d & \longrightarrow & \hat{H}^0(M)^d & \longrightarrow & \hat{H}^0(F)^d & \longrightarrow & \hat{H}^0(N)^d & \longrightarrow 0 \end{array}$$

(the v_i and w_i are given by the cup-products). Now v_3 is an isomorphism, so v_4 is surjective. Since $\mathbf{N}$ is also finitely generated, this result applied to $\mathbf{N}$ shows that v_2 is surjective. The five lemma shows that v_4 is injective; hence, bijective. Once again application to $\mathbf{N}$ gives us the fact that v_2 is an isomorphism, and the five lemma proves that v_1 is an isomorphism. We know w_3 is an isomorphism and we have just shown that w_1 is an isomorphism. Hence, w_4 is surjective while w_2 is injective. But the injectivity of w_2 implies that of w_4 by replacing $\mathbf{M}$ by $\mathbf{N}$; hence, w_4 is an isomorphism. The five-lemma now shows that w_2 is an isomorphism, and we are done. Q.E.D.

§6. Duality Theorem for Finite Group Schemes

We are going to prove the following theorem in this section.

THEOREM 46. *Let* k *be a local field and let* G *be a finite group scheme over* k. *Then the canonical pairing* $G \times G^D \to \mathbf{G}_m$ *yields the cup-product pairings*

$$H^r(k, G) \times H^{2-r}(k, G^D) \to H^2(k, \mathbf{G}_m) = \mathbf{Q}/\mathbf{Z}$$

which are perfect dualities of locally compact groups in the Pontrjagin sense.

The proof of this theorem involves the topology as well as the algebra of the situation; so we must check that the cup-product pairing is continuous. On the cochain level, our product is given by polynomial mappings and is continuous. In view of the extensive results of §4, all that remains to prove the continuity of the cup-product pairing is

PROPOSITION 81. *If* k *is a local field and* K/k *is a finite extension, then* $B^2(K/k, \mathbf{G}_m)$ *is open in* $Z^2(K/k, \mathbf{G}_m)$. *That is,* $H^2(K/k, \mathbf{G}_m)$ *is discrete in its natural topology.*

Proof: In $\otimes^3 K = K \otimes_k K \otimes_k K$ we have the set $\mathfrak{O}_3$ (recall that $\mathfrak{O}_3$ is the set of all $a \in \otimes^3 K$ such that $a = \Sigma_i\, \xi_1^{(i)} \otimes \xi_2^{(i)} \otimes \xi_3^{(i)}$ with $\xi_j^{(i)} \in \mathcal{O}_K$ for all i and j). A fundamental system of neighborhoods of 1 in $(\otimes^3 K)^*$ is given by $\{1 + t^m \mathfrak{O}_3\}_{m=1}^{\infty}$, and we claim that *every 2-cocycle in* $1 + t\mathfrak{O}_3$ *is a coboundary.* Clearly, by establishing this claim we will prove the proposition. But our claim follows from the completeness of K and the statement: *If* $\mathfrak{z}_0$ *is a two cocycle in* $1 + t^m \mathbf{O}_3$*, then there is a cochain* w *in* $1 + t^m \mathbf{O}_2$ *such that* $\mathfrak{z}_0 (\delta w)^{-1} \in 1 + t^{m+1}\mathbf{O}_3$. The latter statement is equivalent to the vanishing of the two dimensional cohomology group of the complex $\{(1 + t^m \mathbf{O}_r)/(1 + t^{m+1}\mathbf{O}_r)\}_{r=0}^{\infty}$; it is this vanishing statement which we shall prove.

The mapping $1 + t^m a \to t^m a$ yields an isomorphism of

$$(1 + t^m \mathbf{O}_r)/(1 + t^{m+1}\mathbf{O}_r)$$

with $t^m \mathbf{O}_r / t^{m+1}\mathbf{O}_r$, where the former group is multiplicative, the latter additive. This mapping commutes with the coboundary operators; so it is an isomorphism of complexes. The groups $\mathbf{O}_r / t\mathbf{O}_r$ and $t^m \mathbf{O}_r / t^{m+1}\mathbf{O}_r$ are naturally (cochain) isomorphic for all values of $m \geqq 0$; so we are reduced to studying the cohomology of $\{\mathbf{O}_r / t\mathbf{O}_r\}_{r=0}^{\infty}$.

Let $1, w_1, \ldots, w_{s-1}$ be an integral basis for $\mathcal{O}_K$ over $\mathcal{O}$, then any element of $\mathbf{O}_{r+1}$ has the form

$$\Sigma\ a_{i_0, \ldots, i_r} w_{i_0} \otimes \cdots \otimes w_{i_r}\,, \quad a_{(i)} \in \mathcal{O}\,,$$

and elements of $t\mathbf{O}_{r+1}$ are similar save that $a_{(i)} \in t\mathcal{O}$. From this, we see that the mapping which sends each $a_{i_0, \ldots, i_r}$ to the residue field $\tilde{k}$ of k yields an isomorphism of the complexes

$$0 \to \mathbf{O}_1/t\mathbf{O}_1 \to \mathbf{O}_2/t\mathbf{O}_2 \to \cdots$$

$$0 \to \tilde{k}^s \to \tilde{k}^{s^2} \to \cdots .$$

A simple computation shows that the last complex has trivial cohomology in positive dimensions, and, as explained above, this completes the proof.

Q.E.D.

Proof of Theorem 46. The proof will have many steps; for ease of exposition we shall label and present them separately.

Step 1. *Reduction to the case of additive and multiplicative type group schemes when* $r = 0, 1, 2$.

We assume the duality true when G is of additive or multiplicative type and $r = 0, 1, 2$. For simplicity of notation, $H^r(k, G)$ will be abbreviated $H^r(G)$. Let G be a connected finite group scheme, then we have the exact sequences

$$0 \longrightarrow G_m \longrightarrow G \longrightarrow G_a \longrightarrow 0$$

$$0 \longrightarrow G_a^D \longrightarrow G^D \longrightarrow G_m^D \longrightarrow 0\,,$$

and the corresponding cohomology sequences are

$$0 \longrightarrow H^1(G_m) \xrightarrow{\theta_1} H^1(G) \xrightarrow{\theta_2} H^1(G_a) \xrightarrow{\delta^*} H^2(G_m) \xrightarrow{\theta_3} H^2(G_0) \longrightarrow 0$$

$$0 \longrightarrow H^0(G^D) \xrightarrow{\phi_1} H^0(G_m^D) \xrightarrow{\delta^{**}} H^1(G_a^D) \xrightarrow{\phi_2} H^1(G^D) \xrightarrow{\phi_3} H^1(G_m^D) \longrightarrow 0$$

(use the Corollary of Proposition 77). Now $H^0(G_m^D)$ is finite; hence δ^{**} is both continuous and open onto its image (which is closed). Therefore, the map $(\delta^{**})^d$ (the superscript "d" denotes Pontrjagin duality) is continuous and the commutative diagram

$$\begin{array}{ccc} H^1(G_a) & \xrightarrow{\delta^*} & H^2(G_m) \\ \downarrow & & \downarrow \\ H^1(G_a^D)^d & \xrightarrow{(\delta^{**})^d} & H^0(G_m^D)^d \end{array}\,,$$

together with our assumption that the vertical maps are topological isomorphisms, shows that δ^* is continuous. Moreover, $\theta_1, \theta_2, \theta_3$ are continuous and open onto their images; the map δ^* is open and closed by discreteness. From these facts, we obtain the commutative diagram

$$\begin{array}{ccccccccccccc} 0 & \to & H^2(G_0)^d & \longrightarrow & H^2(G_m)^d & \longrightarrow & H^1(G_a)^d & \longrightarrow & H^1(G_0)^d & \longrightarrow & H^1(G_m)^d & \longrightarrow & 0 \\ & & \uparrow v_1 & & \uparrow v_2 & & \uparrow v_3 & & \uparrow v_4 & & \uparrow v_5 & & \\ 0 & \to & H^0(G_0^D) & \longrightarrow & H^0(G_m^D) & \longrightarrow & H^1(G_a^D) & \longrightarrow & H^1(G_0^D) & \longrightarrow & H^1(G_m^D) & \longrightarrow & 0\,. \end{array}$$

By assumption, v_2, v_3, v_5 are isomorphisms (topological); the continuity of cup-products, the five lemma, and [P, Theorem 13] imply that v_1, v_4 are also topological isomorphisms. The case $r = 0$ is trivial because both groups vanish.

If G is not connected, we consider the sequence

$$0 \to G_0 \to G \to G_s \to 0 ,$$

and its dual sequence. We may assume $\mathrm{ch}\, k = p > 0$, else all group schemes are multiplicative and our assumption yields the result. When $\mathrm{ch}\, k = p > 0$, we again may assume $|G| =$ power of p, else we are back in the multiplicative case. Now the cohomology sequence is

$$0 \to H^0(G) \xrightarrow{\theta_1} H^0(G_s) \xrightarrow{\delta^*} H^1(G_0) \xrightarrow{\theta_2} H^1(G) \xrightarrow{\theta_3} H^1(G_s) \xrightarrow{\delta^{**}}$$
$$\to H^2(G_0) \xrightarrow{\theta_4} H^2(G) \longrightarrow 0 \;;$$

by discreteness, δ^{**} is continuous and by local compactness the continuous maps θ_i $(i = 1,2,3,4)$ are open onto their images and each image is closed. Moreover, δ^* is continuous and is a closed map. Just as above, we obtain the commutative diagram

$$\begin{array}{ccccccccccccccccc}
0 \to & H^2(G)^d & \to & H^2(G_0)^d & \to & H^1(G_s)^d & \to & H^1(G)^d & \to & H^1(G_0)^d & \to & H^0(G_s)^d & \to & H^0(G)^d & \to & 0 \\
& \uparrow {\scriptstyle v_1} & & \uparrow {\scriptstyle v_2} & & \uparrow {\scriptstyle v_3} & & \uparrow {\scriptstyle v_4} & & \uparrow {\scriptstyle v_5} & & \uparrow {\scriptstyle v_6} & & \uparrow {\scriptstyle v_7} & & \\
0 \to & H^0(G^D) & \to & H^0(G_0^D) & \to & H^1(G_s^D) & \to & H^1(G^D) & \to & H^1(G_0^D) & \to & H^2(G_s^D) & \to & H^2(G^D) & \to & 0 \;.
\end{array}$$

The maps v_3, v_6 are topological isomorphisms by assumption, the maps v_2, v_5 are topological isomorphisms by the first part of this argument; so we deduce from the 5-lemma and [P, Theorem 13] that v_1, v_4, v_7 are also topological isomorphisms.

Step 2. *The cohomology* $H^r(k, G)$ *vanishes for* $r \geqq 3$. By Step 1, $H^2(k, G)$

is dual to $H^0(k, G^D)$; so it follows that H^2 is right exact on the abelian category of finite group schemes over k. We shall prove $H^r(k, G) = (0)$ for $r \geqq 3$ by induction on r. Observe that the zero functor being right-exact, our result will follow from the statement: *If* $H^i(-)$ *is right-exact on the finite group schemes, then* $H^{i+1}(-)$ *vanishes on the finite group schemes.*

Let G be a finite group scheme, and choose any $\xi \in H^{i+1}(G)$. We choose a finite field extension K/k such that $\mathrm{res}(\xi) \in H^{i+1}(K, G \otimes K)$ vanishes (cf. the proof of Theorem 42), and we form the finite group scheme $\pi_*(K \to k)(G \otimes K)$. Then the exact sequence,

$$0 \longrightarrow G \xrightarrow{j} \pi_*(K \to k)(G \otimes K) \longrightarrow G'' \longrightarrow 0$$

yields the commutative diagram

$$\begin{array}{ccc} H^{i+1}(G) & \xrightarrow{\quad \mathrm{res} \quad} & H^{i+1}(K, G \otimes K) \\ & {\scriptstyle j_*} \searrow & \uparrow {\scriptstyle sh_{i+1}} \\ & & H^{i+1}(k, \pi_*(K \to k)(G \otimes K)) \end{array} .$$

It follows that $j_*(\xi) = 0$; but H^i is right-exact, so j_* is a monomorphism and ξ vanishes.

Step 3. *Duality in the additive case when* $r = 0, 1, 2$. According to the corollary of Proposition 77, the only non-trivial case is when $r = 1$. In this case, we use induction on the length of a composition series for the group scheme G. This goes as follows: Let

$$0 \to G' \to G \to G'' \to 0$$

be an exact sequence of additive type group schemes, and consider its dual sequence as well. We obtain cohomology sequences

$$0 \longrightarrow H^1(G') \xrightarrow{\theta_1} H^1(G) \xrightarrow{\theta_2} H^1(G'') \longrightarrow 0$$

$$0 \longrightarrow H^1(G''^D) \xrightarrow{\theta_3} H^1(G^D) \xrightarrow{\theta_4} H^1(G'^D) \longrightarrow 0 .$$

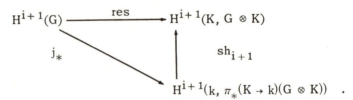

The usual arguments (see Step 1) show that each θ_i is continuous, open onto its image, and each image is closed. Pontrjagin duality theory yields the usual commutative diagram

$$\begin{array}{ccccccccc} 0 & \longrightarrow & H^1(G') & \longrightarrow & H^1(G) & \longrightarrow & H^1(G'') & \longrightarrow & 0 \\ & & \downarrow{\scriptstyle v_1} & & \downarrow{\scriptstyle v_2} & & \downarrow{\scriptstyle v_3} & & \\ 0 & \longrightarrow & H^1(G'^D)^d & \longrightarrow & H^1(G^D)^d & \longrightarrow & H^1(G''^D)^d & \longrightarrow & 0 \end{array},$$

and the five-lemma (together with [P, Theorem 13] and continuity of the cup-product) completes the induction.

According to Proposition 77, the essential case is then $G = \alpha_p$, $G^D = \alpha_p$. We know that $H^1(\alpha_p) = k^+/k^{+p}$, and this is true topologically as well. (For Proposition 76 shows that $H^1(k_1/k, \alpha_p)$ is topologically isomorphic to $H^1(\alpha_p)$. Since $H^1(k_1/k, G_a) = (0)$, we deduce that $H^1(k_1/k, \alpha_p)$ is topologically isomorphic to $k_1{}^+/k^+$, i.e., to k^+/k^{+p} *via* raising to the pth power.) The duality pairing $\alpha_p \times \alpha_p \to G_m$ is given explicitly by

$$P(T) = \sum_{i=0}^{p-1} \left(\frac{1}{i!}\right) y^i \otimes z^i = \exp(y \otimes z) ,$$

where $\alpha_p = \operatorname{Spec} k[y]/(y^p) = \operatorname{Spec} k[z]/(z^p)$ and $G_m = \operatorname{Spec} k[T, T^{-1}]$. Also, exp, refers to the "truncated" exponential function. The cup product formula of §3 shows that the cocycle induced in $Z^2(k_1/k, G_m)$ by the pairing P is given by

$$(*) \qquad \underline{P}(f, g) = \exp(cd\delta_n \varepsilon_m) ,$$

whenever $f = at^n$, $g = bt^m$; where $c^p = a$, $d^p = b$, $a, b \in \tilde{k}$, and

$$\delta_n = 1 \otimes \theta_1{}^n \otimes 1 - \theta_1{}^n \otimes 1 \otimes 1 \in k_1 \otimes k_1 \otimes k_1$$

$$\varepsilon_m = (1 \otimes 1 \otimes \theta_1{}^m) - 1 \otimes \theta_1{}^m \otimes 1 \in k_1 \otimes k_1 \otimes k_1 .$$

Let $cl(\underline{P}(f, g))$ denote the cohomology class of $\underline{P}(f, g)$ considered in the group $H^2(k, G_m)$, and define the symbol (f, g) by the usual formula:

$(f, g) = \operatorname{inv}_k(\mathrm{cl}(\underline{P}(f, g)))$. Then the symbol (f, g) represents the pairing of k^+/k^{+p} with itself into $\mathbf{Q}/\mathbf{Z}$ induced by the cup-product from Cartier duality. It is not hard to see that Proposition 81 and the explicit formula for $\underline{P}$ allow one to prove all statements concerning the pairing from the special case $f = at^n$, $g = bt^m$. Clearly, $(at^n, bt^m) = (abt^n, t^m)$, so we may even assume $b = 1$. Computations involving (*) show that the following formulas are valid:

$$(at^n, t^m) = -(at^m, t^n)$$

$$(at^n, t^m) = m(at^{n+m-1}, t) \ .$$

For example, (*) implies that

$$\underline{P}(at^n, t^m)\,\underline{P}(at^m, t^n) = \delta(\exp(-c\gamma_n\gamma_m))$$

where $c^p = a$, $\gamma_n = 1 \otimes \theta_1^n - \theta_1^n \otimes 1$. This statement yields the first formula above, and the second follows from similar computations. Thus, determination of the symbol (f, g) is reduced to the case $f = at^n$, $g = t$.

In this case, the two formulas above imply that $(at^n, t) = 0$ if $n \not\equiv -1 \pmod p$, while if $n \geqq p$, Proposition 81 shows that the symbol vanishes. Explicit computations in the cases $n \neq -1$, show that the symbol also vanishes, so only the case $n = -1$ remains. Let K be the unramified extension of k obtained by adjoining to k the roots of the equation $\wp(x) = x^p - x = a$; and let σ denote the Frobenius automorphism of K/k. The cocycle $z(\sigma^i, \sigma^j)$ given by

$$z(\sigma^i, \sigma^j) = 1, \quad i + j < p; \quad z(\sigma^i, \sigma^j) = t^{\mathrm{tr}(a)}, \quad i + j \geqq p,$$

where $\mathrm{tr}(a)$ means the trace from $\tilde{k}$ to $\mathbf{Z}/p\mathbf{Z}$ of a, is cohomologous to $\underline{P}(at^{-1}, t)$. (Inflate both to the compositum of K and k_1 over k.) But the invariant of the cocycle z is exactly $1/p \cdot \mathrm{tr}(a)$, as one sees from the explicit form of inv_k in unramified extensions (Proposition 51). This shows that the symbol (f, g) is precisely $\frac{1}{p}\,\mathrm{tr}(\mathrm{res}(f\,dg))$, and it is well-known that the formula

$$<f, g> \longmapsto \mathrm{tr}(\mathrm{res}(f\,d\,g)) \in \mathbf{Z}/p\mathbf{Z}$$

places k^+/k^{+p} in exact Pontrjagin duality with itself.

Step 4. *Decomposition in the multiplicative case.* If G is a group scheme (finite over k) of multiplicative type, then G is the direct product of its p-torsion subgroup scheme and its prime-to-p subgroup scheme. One sees this by considering the *étale* group scheme, G^D, recalling that G^D corresponds to a finite Galois module, and applying ordinary abelian group theory. It follows that the argument for multiplicative type group schemes is reduced to the two cases: prime-to-p, p-torsion.

Step 5. *Multiplicative type, prime-to-*p. There are two proofs in this case, the first (due to Serre) is very slick and uses dualizing modules and Theorem 44. It goes as follows: Since G and G^D are both *étale,* and since $I = \mu = \operatorname{dir\,lim}_n \mu_n$ by Theorem 44, we see that in the notation of §1, $G^{(d)} = G^D$. This being said, the cases $r = 0, 2$ of our theorem follow instantly from the Corollary to Proposition 65 if one uses the interchangeability of G and G^D (*both* are *étale*!). For the case $r = 1$, the interchangeability of G and G^D shows that only injectivity need be proved for the map $H^1(k, G) \to H^1(k, G^D)^d$. If $\xi \in H^1(k, G)$, there exists a finite extension K/k such that $\mathrm{res}(\xi) = 0$ i.e., such that the monomorphism of schemes

$$0 \to G \xrightarrow{g} \pi_*(K \to k)\, G$$

effaces the cohomology class ξ. If $C = \operatorname{coker} g$, then the exact sequence of cohomology yields the commutative diagram

$$\begin{array}{ccccccc}
H^0(\pi_*(K \to k)\, G) & \longrightarrow & H^0(C) & \xrightarrow{\delta} & H^1(G) & \xrightarrow{g_*} & H^1(\pi_*(K \to k)\, G) \\
\downarrow v_1 & & \downarrow v_2 & & \downarrow v_3 & & \\
H^2(\pi_*(K \to k)\, G^D)^d & \longrightarrow & H^2(G^D)^d & \longrightarrow & H^1(G^D)^d & & .
\end{array}$$

Now $g_*(\xi) = 0$, so $\xi \in \operatorname{Im} \delta$. Since v_1, v_2 are bijective (by the above $r = 0$ case), elementary diagram chasing shows that if $v_3(\xi) = 0$, then $\xi = 0$.

The second proof is due to Tate, it uses Theorem 45. We are given G of multiplicative type, so its dual, G^D, is *étale* and we have the exact sequence

$$(*)\qquad 0 \to N \to M \to G^D \to 0$$

with M, N torsion-free modules of the same (finite) $\mathbb{Z}$-rank. The dual sequence is then

$$(**)\qquad 0 \to G \to X_M \to X_N \to 0 ,$$

and we obtain the following cohomology sequences

$$0 \to \hat{H}^0(N) \to \hat{H}^0(M) \to H^0(G^D) \to H^1(N) \to H^1(M) \to H^1(G^D) \to H^2(N) \to H^2(M) \to H^2(G^D) \to 0$$

$$0 \to H^0(G) \to \hat{H}^0(X_M) \to \hat{H}^0(X_N) \to H^1(G) \to H^1(X_M) \to H^1(X_N) \to H^2(G) \to H^2(X_M) \to H^2(X_N) \to 0.$$

There is no trouble in passing to completions since $H^0(G^D)$ and $H^1(G)$ are finite, G being *étale* as well as multiplicative. Moreover, continuous (= torsion) homomorphisms into $\mathbb{Q}/\mathbb{Z}$ is the Pontrjagin dualizing functor, so we may pass to the dual sequences at will. We get the commutative diagram

$$\begin{array}{ccccccccccccccccccccc}
0 \to & \hat{H}^0(N) & \to & \hat{H}^0(M) & \to & H^0(G^D) & \to & H^1(N) & \to & H^1(M) & \to & H^1(G^D) & \to & H^2(N) & \to & H^2(M) & \to & H^2(G^D) & \to 0 \\
& \downarrow{\scriptstyle v_1} & & \downarrow{\scriptstyle v_2} & & \downarrow{\scriptstyle v_3} & & \downarrow{\scriptstyle v_4} & & \downarrow{\scriptstyle v_5} & & \downarrow{\scriptstyle v_6} & & \downarrow{\scriptstyle v_7} & & \downarrow{\scriptstyle v_8} & & \downarrow{\scriptstyle v_9} & \\
0 \to & H^2(X_N)^d & \to & H^2(X_M)^d & \to & H^2(G)^d & \to & H^1(X_N)^d & \to & H^1(X_M)^d & \to & H^1(G)^d & \to & \hat{H}^0(X_N)^d & \to & \hat{H}^0(X_M)^d & \to & H^0(G)^d & \to 0.
\end{array}$$

Theorem 45 shows that $v_1, v_2, v_4, v_5, v_7, v_8$ are isomorphisms; so the five lemma implies that v_3, v_6, v_9 are isomorphisms, as required.

Step 6. *Multiplicative type, p-torsion.* In this case G^D is *étale*, but G is not; so there is no interchangeability possible. Moreover, $H^1(G^D)$

and $H^1(G)$ are infinite groups with a non-trivial topology so that in using Theorem 45 care must be exercised as to the continuity and closedness of maps. Actually all this can be done, but we prefer to proceed by going around these difficulties rather than through them.

Once again as in the second proof of Step 5 we have exact sequences (*) and (**). Note that the duality between $H^0(G)$ and $H^2(G^D)$ is trivial as both groups vanish. We treat the case of $H^2(G)$ and $H^0(G^D)$ first. By the cohomology sequences, we obtain the commutative diagram

$$\begin{array}{ccccccccc} H^1(X_M) & \to & H^1(X_N) & \to & H^2(G) & \to & H^2(X_M) & \to & H^2(X_N) \\ \downarrow & & \downarrow & & \downarrow & & \downarrow & & \downarrow \\ H^1(M)^d & \to & H^1(N)^d & \to & H^0(G^D)^d & \to & \hat{H}^0(M)^d & \to & \hat{H}^0(N)^d \end{array}$$

so the five-lemma settles the issue.

Only the duality of $H^1(G)$ and $H^1(G^D)$ remains, and this is where care must be exercised. The sequence (**) yields the cohomology sequence

$$0 \to H^0(X_M) \to H^0(X_N) \to H^1(G) \to H^1(X_M) \to H^1(X_N) \,;$$

and if $\mathbf{D}$ denotes the functor $\mathrm{Hom}(-, \mathbf{R}/\mathbf{Z})$ (no topology), then the diagram

$$\begin{array}{ccccccccc} \mathbf{D}H^1(X_N) & \to & \mathbf{D}H^1(X_M) & \to & \mathbf{D}H^1(G) & \to & \mathbf{D}H^0(X_N) & \to & \mathbf{D}H^0(X_M) \\ \uparrow{\scriptstyle w_1} & & \uparrow{\scriptstyle w_2} & & \uparrow{\scriptstyle w_3} & & \uparrow{\scriptstyle w_4} & & \uparrow{\scriptstyle w_5} \\ H^1(N) & \to & H^1(M) & \to & H^1(G^D) & \to & H^2(N) & \to & H^2(M) \end{array}$$

commutes. Theorem 45 shows that w_4, w_5 are injections; so the five-lemma implies that w_3 is injective. However, cup-product is continuous; and it follows that w_3 is the composition

$$H^1(G^D) \xrightarrow{\ v\ } H^1(G)^d \hookrightarrow \mathbf{D}H^1(G) \ .$$

This proves that v is injective, and only its surjectivity remains to be proved. I claim that if this is known when G^D is a module with trivial action, then it follows in general. For if K is a trivializing field for G and G^D, then the inflation-restriction sequence yields the injection of compact groups

$$0 \to H^1(k, G) \xrightarrow{\text{res}} H^1(K, G \otimes K) \ ;$$

and, by duality, the surjection of discrete groups

$$H^1(K, G \otimes K)^d \xrightarrow{\text{res}^d} H^1(k, G)^d \longrightarrow 0 \ .$$

However, tr is dual to res under cup-products; so the diagram

$$\begin{array}{ccccc} H^1(K, (G \otimes K)^D) & \xrightarrow{\text{tr}} & H^1(k, G^D) & & \\ \downarrow{\scriptstyle v_K} & & \downarrow{\scriptstyle v_k} & & \\ H^1(K, G \otimes K)^d & \xrightarrow{\text{res}^d} & H^1(k, G)^d & \longrightarrow & 0 \end{array}$$

is commutative. Since we assume v_K is surjective, it follows that v_k is also surjective.

We finally have to consider only those G whose duals are modules with trivial action. By the same arguments as in Step 3, we reduce the issue to the essential case: $G = \mu_p$, $G^D = \mathbf{Z}/p\mathbf{Z}$. Now the exact sequence

$$0 \longrightarrow \mathbf{Z}/p\mathbf{Z} \longrightarrow \mathbf{G}_a \xrightarrow{\wp} \mathbf{G}_a \longrightarrow 0$$

where $\wp(x) = x^p - x$, shows that $H^1(k, \mathbf{Z}/p\mathbf{Z}) = k^+/\wp(k^+)$. It also shows that $\mathbf{Z}/p\mathbf{Z}$ is Spec $k[x]/(x^p - x)$. The group scheme μ_p is Spec $k[y]/(y^p - 1)$ and the duality is set up as follows: A basis for the algebra of $\mathbf{Z}/p\mathbf{Z}$ is $1, x, \dots, x^{p-1}$; let its dual basis be $f_0, f_1, \dots, f_{p-1}$. One can easily check the following facts: The f_i are divided powers of f_1, i.e., $f_i = f_1^i/i!$, $y = \exp(f_1) = f_0 + f_1 + f_2 + \cdots + f_{p-1}$. Hence, the duality pairing $G \times G^D \to \mathbf{G}_m$ is given by

$$P(T) = \exp(x \otimes \log y) \ .$$

Precisely the same arguments and computations as in Step 3 show that the induced cup-product pairing

$$k^*/k^{*p} = H^1(\mu_p) \times H^1(Z/pZ) = k^+/\wp(k^+) \to Q/Z$$

is given by

$$<\xi, \eta> \longmapsto \frac{1}{p} \operatorname{tr}(\operatorname{res} \eta \frac{d\xi}{\xi}) \quad ;$$

so that Chapter V now completes the proof. Q.E.D.

Remarks 1. It is clear now how the various local class field theory symbols fit together as cup-products in the duality of Theorem 46.

2. If ch $k = 0$, or if one stays away from p-torsion in char $p > 0$, then Theorem 44 and Step 5 are all that is necessary for the proof of Theorem 46. This is the case originally studied by Tate, [Lg, TD].

3. One can put together Theorems 45 and 46 to obtain a duality theorem for those group schemes X which fit into an exact sequence of the type

$$0 \to G \to X \to X_M \to 0$$

or of the type

$$0 \to X_M \to X \to G \to 0$$

where G is a finite group scheme. We leave details of this to the reader. Actually, Tate proves [TD] a duality theorem for abelian varieties over k and one can reformulate all these results as a vast duality theorem of "complexes of group schemes" as Grothendieck has suggested. This is the point of view taken in the current literature.

BIBLIOGRAPHY

[A] E. Artin, "Galois Theory," Notre Dame Lecture Notes, No. 2, 1957.

[AG] M. Artin, "Grothendieck Topologies," Mimeographed notes at Harvard University, 1962.

[Ax] J. Ax, "A field of cohomological dimension 1 which is not C_1," Bull. Amer. Math. Soc. 71 (1965), p. 717.

[AK] J. Ax, S. Kochen, "Diophantine Problems over Local Fields I, II," Amer. J. Math. 87 (1965), pp. 605-631-649; III, Ann. of Math. 83 (1966), pp. 437-456.

[Bi] B. J. Birch, "Homogeneous forms of odd degree in a large number of variables," Mathematika 4 (1957), pp. 102-105.

[Br] R. Brauer, "Über Systeme hyperkomplexer Zahlen," Math. Zeit. 30 (1929), pp. 79-107.

[B] D. Buchsbaum, "Satellites and Universal Functors," Ann. of Math. 71 (1960), pp. 199-209.

[CE] H. Cartan, S. Eilenberg, *Homological Algebra*, Princeton Univ. Press, Princeton, N. J., 1956.

[D] A. Douady, "Cohomologie des groups compacts totalement discontinus," Sem. Bourb. Éxp. 189, 1959-60.

[Dk] B. Dwork, "Norm residue symbol in local number fields," Hamb. Abh. 22 (1958), pp. 180-190.

[EM] S. Eilenberg, J. Moore, "Limits and Spectral Sequences," Topology 1 (1962), pp. 1-23.

[Ga] P. Gabriel, "Sur les hyperalgébras étudiees par J. Dieudonné," Sem. J.-P. Serre, Paris, 1960.

[Go] R. Godement, *Topologie Algébrique et Théorie des Faisceaux*, Herman et Cie, Paris, 1958.

[Gg] M. Greenberg, *Lectures on Forms in Many Variables*, Benjamin, New York, 1969.

[GD] A. Grothendieck, J. Dieudonné, *Éléments de Géométrie Algébrique III*, I.H.E.S. Pub. Math. #11, Paris, 1961.

[GD_2] ____________, Op. Cit. IV.

[GDm] A. Grothendieck, M. Demazure, Schemas en Groupes, 1, 2, 3, Lecture Notes in Math. Vols. 151, 152, 153, Springer-Verlag, New York, 1970.

[GG] A. Grothendieck, "Technique de descent et Théorèmes d'Existence en Géométrie Algébrique, I," Sem. Bourb. Éxp. 190, 1959-60.

[GF] ____________, Op. Cit. II, Sem. Bourb. Éxp. 195, 1959-60.

[GQ] ____________, Op. Cit. III, Sem. Bourb. Éxp. 212, 1960-61.

[GH] ____________, Op. Cit. IV, Sem. Bourb. Éxp. 221, 1960-61.

[GB] ____________, "Groupe de Brauer, III," Sem. d'I.H.E.S. Paris, 1968.

[GT] ____________, "Sur Quelques points d'Algèbre Homologique," Tohoku Math. J. 9 (1957), pp. 119-221.

[Ha] H. Hasse, *Zahlentheorie*, Akademie-Verlag, Berlin, 1949.

[He] E. Hecke, *Vorlesungen über die Theorie der algebraischen Zahlen*, Akademische Verlag, Leipzig, 1923.

[Hl] D. Hilbert, *Die Theorie der algebraischen Zahlkörper*, Jahresbericht D. Math. Ver. Bd. 4 (1897), pp. 175-546.

[HS] G. Hochschild, J.-P. Serre, "Cohomology of Group Extensions," Trans. Amer. Math. Soc. 74 (1953), pp. 110-134.

[HJ] K. Hoechsmann, J. Gamst, "Products in Sheaf Cohomology," Tohoku Math. J., 22, (1970), pp. 143-162.

[I] K. Iwasawa, "A note on the group of units of an algebraic number field," J. Math. Pures et Appl. 35 (1956), pp. 189-192.

[K] D. Kan, "Adjoint Functors," Trans. Amer. Math. Soc. 87 (1958), pp. 294-329.

[LA] S. Lang, *Algebra*, Addison-Wesley, Reading, Mass., 1965.

[LD] ______________, *Diophantine Geometry*, Interscience Publishers, New York, 1962.

[LQ] ______________, "On Quasi-Algebraic closure," Ann. of Math. 55 (1952), pp. 373-390.

[LR] ______________, *Rapport sur la cohomologie des groupes*, Benjamin, New York, 1966.

[Lg] ______________, "Cohomology of Abelian varieties over p-adic fields" (after J. Tate), mimeographed notes at Princeton Univ., 1959.

[LT] J. Lubin, J. Tate, "Formal complex multiplication in local fields," Ann. of Math. 81 (1965), pp. 380-387.

[ML] D. Mumford, Lectures on Curves on an Algebraic Surface, Ann. of Math. Studies 59, Princeton Univ. Press, Princeton, N. J., 1966.

[N] T. Nakayama, G. Hochschild, "Cohomology in Class Field Theory," Ann. of Math. 55 (1952), pp. 348-366.

[O] F. Oort, *Commutative Group Schemes*, Lecture notes in Math 15, Springer-Verlag, Berlin, 1966.

[Pe] L. G. Peck, "Diophantine equations in algebraic number fields," Amer. J. Math. 71 (1949), pp. 387-402.

[P] L. Pontrjagin, *Topological Groups,* Princeton University Press, Princeton, N. J., 1946.

[R] P. Roquette, "On Class Field Towers," Alg. No. Thy. (Proc. Brighton Conf.) Thompson Book Co., Washington, D. C., 1967, pp. 231-249.

[RZ] A. Rosenberg, D. Zelinsky, "On Amitsur's Complex," Trans. Amer. Math. Soc. 97 (1960), pp. 327-356.

[S] I. Šafarevič, "Algebraic Number Fields" (in Russian), Proc. Stockholm Cong., 1962, pp. 163-176.

[SG] I. Šafarevič, E. Golod, "On Class Field Towers" (in Russian), Izv. Akad. Nauk. S.S.S.R. 28 (1964), pp. 261-272.

[Sd] H. Schmid, "Über das Reziprozitätsgesetz in relativzyklischen algebraischen Funktionenkörpern mit endlichen Konstenten-körper," Math. Zeit 40 (1936), pp. 94-109.

[SA] J.-P. Serre, *Groupes Algébriques et Corps des Classes,* Hermann et Cie, Paris, 1959.

[SCL] ____________, *Corps Locaux,* Hermann et Cie, Paris, 1962.

[SG] ____________, *Cohomologie Galoisienne,* Lecture Notes in Math, no. 5, Springer-Verlag, Berlin, 1964.

[SSS] S. Shatz, "The cohomological dimension of certain Grothendieck topologies," Ann. of Math. 83 (1966), pp. 572-595.

[SSh] ____________, "The structure of the category of sheaves in the flat topology over certain local rings," Amer. J. Math. 90 (1968), pp. 1346-1354.

[TC] J. Tate, "The higher dimensional cohomology groups of class field theory," Ann. of Math. 56 (1952), pp. 294-297.

[TD] ____________, "Duality Theorems in Galois cohomology over number fields," Proc. Stockholm Cong., 1962, pp. 288-295.

[TJ] G. Terjanian, "Un contre-example à une conjecture d'Artin," C. R. Acad. Sci. Paris 262 (1966), p. 612.

[Wg] E. Warning, "Bermerkung zur vorstehenden Arbeit von Herrn Chevalley," Abh. Math. Sem. Univ. Hamburg 11 (1936), pp. 76-83.

[Wl] A. Weil, *Foundations of Algebraic Geometry,* Amer. Math. Soc. Coll. Pub. 29, Providence, 1946.

[W] E. Witt, "Zyklische Körper und Algebren der Charakteristik p vom Grade p^n," J. Reine und Ang. Math. 176 (1936), pp. 126-140.